T0255110

UNITEXT - La Matematica per il 3+2

Volume 102

More information about this series at http://www.springer.com/series/5418

Carlo Viola

An Introduction to Special Functions

Springer

Carlo Viola
Department of Mathematics
University of Pisa
Pisa
Italy

ISSN 2038-5722 ISSN 2038-5757 (electronic)
UNITEXT - La Matematica per il 3+2
ISBN 978-3-319-41344-0 ISBN 978-3-319-41345-7 (eBook)
DOI 10.1007/978-3-319-41345-7

Library of Congress Control Number: 2016944325

Mathematics Subject Classification (2010): 30-01, 30A10, 33B15, 33C05, 33C15, 34M03, 65B15

Printed on acid-free paper

This Springer imprint is published by Springer Nature
The registered company is Springer International Publishing AG Switzerland

Preface

These lecture notes stem from a course that I gave at the doctoral school in mathematics at Pisa University during the academic year 2013–2014. Their primary purpose is to expound some material well suited for a second course on analytic functions of one complex variable, after a first elementary course dealing with the basic concepts in this theory, such as the residue theorem, Cauchy's integral formula, Taylor and Laurent series expansions, poles and essential singularities, branch points, etc. These basic subjects are the only background assumed in this book; I have made a serious attempt to avoid more advanced prerequisites, sometimes at the cost of choosing slightly longer but more elementary proofs of the theorems.

As the title suggests, however, the topics included have been especially chosen to provide the reader with the main notions and results in the theory of functions of one complex variable leading to a rigorous treatment of some special functions: in the first place, the Euler gamma function, which can be considered the most frequently used non-elementary mathematical function, and, secondly, the most important in the family of hypergeometric functions, namely the Euler–Gauss hypergeometric function ${}_2F_1$ and the Kummer confluent hypergeometric function ${}_1F_1$. These functions are indispensable tools in 'higher calculus' and are often encountered in all branches of pure and applied mathematics. I have tried to treat, on a solid ground, the basic material concerning these functions in the hope of providing a reasonably detailed exposition of their main properties.

The content of these notes is classical, and therefore there is nothing original, except perhaps the selection of the subject matter. Naturally, I have borrowed much from many excellent books, some of which are listed in the bibliography, and it would be impossible to acknowledge in detail all my indebtedness to other authors.

Special thanks are due to A. Perelli, who kindly read a preliminary draft of these lecture notes and made valuable remarks and suggestions.

Pisa, Italy — Carlo Viola
May 2016

Contents

Chapter 1
Picard's Theorems

Our main goal in this chapter is to give an elementary proof of Picard's first and second theorems, which we base upon Schottky's theorem (Theorem 1.2). We begin with a proof of Borel–Carathéodory's inequalities (1.1) and (1.2), which yield useful upper bounds for $|f(z)|$ and $\left|f^{(n)}(z)\right|$ $(n \geq 1)$ in terms of $\operatorname{Re} f(z)$. In the proof of Schottky's Theorem 1.2 we shall employ the inequality (1.1), while (1.2) will be used in Chap. 3, in the proof of Hadamard's Theorem 3.4.

1.1 Borel–Carathéodory's Theorem

Theorem 1.1 (Borel–Carathéodory) *Let $0 < r < R$, let $z_0 \in \mathbb{C}$, and let $f(z)$ be a function regular*[1] *in the closed disc $|z - z_0| \leq R$. Then*

$$\max_{|z-z_0|=r} |f(z)| \leq \frac{2r}{R-r} \max_{|z-z_0|=R} \operatorname{Re} f(z) + \frac{R+r}{R-r} |f(z_0)|. \tag{1.1}$$

A similar upper bound holds for the derivatives of $f(z)$: if

$$\max_{|z-z_0|=R} \operatorname{Re} f(z) \geq 0,$$

[1]Throughout these notes, a function $f(z)$ of a complex variable z is said to be *regular* at a point $z_* \in \mathbb{C}$ if it is holomorphic (i.e., satisfies the Cauchy–Riemann equations) in an open neighbourhood of z_*. The function $f(z)$ is regular in a set $S \subset \mathbb{C}$ if it is regular at every point of S.

C. Viola, *An Introduction to Special Functions*,
UNITEXT - La Matematica per il 3+2 102, DOI 10.1007/978-3-319-41345-7_1

then, for $n = 1, 2, \ldots,$

$$\max_{|z-z_0|=r} \left|f^{(n)}(z)\right| \le \frac{2^{n+2} n!\, R}{(R-r)^{n+1}} \left(\max_{|z-z_0|=R} \operatorname{Re} f(z) + |f(z_0)| \right). \tag{1.2}$$

Proof By a translation, we may plainly assume $z_0 = 0$. The theorem is obviously true if $f(z)$ is constant. Otherwise, we denote for brevity

$$A = \max_{|z|=R} \operatorname{Re} f(z),$$

and we first assume $f(0) = 0$. Since $e^{f(z)}$ is regular in $|z| \le R$, the maximum of $\left|e^{f(z)}\right| = e^{\operatorname{Re} f(z)}$ in $|z| \le R$ is attained only on the border $|z| = R$. Therefore

$$A > \operatorname{Re} f(0) = 0.$$

The function

$$\varphi(z) := \frac{f(z)}{2A - f(z)}$$

is regular in $|z| \le R$ because the real part of the denominator is >0. Let $f(z) = u + iv$ with $u, v \in \mathbb{R}$. From $-2A + u \le u \le 2A - u$ we get

$$|\varphi(z)|^2 = \frac{u^2 + v^2}{(2A-u)^2 + v^2} \le 1.$$

Since $\varphi(0) = f(0) = 0$, the function $\varphi(z)/z$ is regular in $|z| \le R$, whence

$$\left|\frac{\varphi(z)}{z}\right| \le \max_{|z|=R} \left|\frac{\varphi(z)}{z}\right| \le \frac{1}{R}.$$

In particular, for $|z| = r$,

$$\max_{|z|=r} |\varphi(z)| \le \frac{r}{R}.$$

Hence for any z with $|z| = r$ we get

$$|f(z)| = 2A \frac{|\varphi(z)|}{|1+\varphi(z)|} \le 2A \frac{r/R}{1 - r/R} = \frac{2Ar}{R-r}, \tag{1.3}$$

that is (1.1) in the present case $f(0) = 0$.

If $f(0) \ne 0$, we apply (1.3) to the function $f(z) - f(0)$. We obtain

$$|f(z) - f(0)| \le \frac{2r}{R-r} \max_{|z|=R} \operatorname{Re}(f(z) - f(0)) = \frac{2r}{R-r} \left(\max_{|z|=R} \operatorname{Re} f(z) - \operatorname{Re} f(0) \right).$$

Therefore, for $|z| = r$,

$$|f(z)| - |f(0)| \le |f(z) - f(0)| \le \frac{2r}{R-r}\Big(\max_{|z|=R} \operatorname{Re} f(z) + |f(0)|\Big),$$

and (1.1) follows.

If $\max_{|z|=R} \operatorname{Re} f(z) \ge 0$, (1.1) yields

$$\max_{|z|=r} |f(z)| \le \frac{R+r}{R-r}\Big(\max_{|z|=R} \operatorname{Re} f(z) + |f(0)|\Big). \tag{1.4}$$

For any z such that $|z| = r$, by Cauchy's integral formula we have

$$f^{(n)}(z) = \frac{n!}{2\pi i} \oint_{|t-z|=(R-r)/2} \frac{f(t)}{(t-z)^{n+1}}\,\mathrm{d}t. \tag{1.5}$$

Since $|t| \le |z| + |t-z| = r + (R-r)/2 = (R+r)/2$, (1.4) yields

$$\begin{aligned} |f(t)| &\le \max_{|\tau|=(R+r)/2} |f(\tau)| \le \frac{R+\frac{1}{2}(R+r)}{R-\frac{1}{2}(R+r)}\Big(\max_{|\tau|=R} \operatorname{Re} f(\tau) + |f(0)|\Big) \\ &< \frac{4R}{R-r}\Big(\max_{|\tau|=R} \operatorname{Re} f(\tau) + |f(0)|\Big). \end{aligned}$$

Thus, by (1.5),

$$\begin{aligned} \big|f^{(n)}(z)\big| &\le \frac{n!}{((R-r)/2)^n}\,\frac{4R}{R-r}\Big(\max_{|\tau|=R} \operatorname{Re} f(\tau) + |f(0)|\Big) \\ &= \frac{2^{n+2} n!\, R}{(R-r)^{n+1}}\Big(\max_{|\tau|=R} \operatorname{Re} f(\tau) + |f(0)|\Big). \end{aligned}$$

□

1.2 Schottky's Theorem

Lemma 1.1 *Let R_0 and M be positive constants. Let $\varphi(\varrho)$ be a real function such that*

$$0 \le \varphi(\varrho) \le M \tag{1.6}$$

for $0 < \varrho < R_0$. Assume there exists a constant $C > 0$ such that

$$\varphi(\varrho) \le \frac{C}{(R-\varrho)^2}\sqrt{\varphi(R)} \tag{1.7}$$

for any ϱ, R *satisfying* $0 < \varrho < R < R_0$. *Then, for any* ϱ *such that* $0 < \varrho < R_0$,

$$\varphi(\varrho) \leq \frac{4^4 C^2}{(R_0 - \varrho)^4}.$$

Proof Let $0 < \varrho < \varrho_1 < R_0$. From (1.6) and (1.7) with $R = \varrho_1$ we get

$$\varphi(\varrho) \leq \frac{C}{(\varrho_1 - \varrho)^2}\sqrt{\varphi(\varrho_1)} \leq \frac{C}{(\varrho_1 - \varrho)^2} M^{1/2}.$$

Thus, for $0 < \varrho < \varrho_1 < \varrho_2 < R_0$,

$$\varphi(\varrho_1) \leq \frac{C}{(\varrho_2 - \varrho_1)^2} M^{1/2},$$

whence, again by (1.7) with $R = \varrho_1$,

$$\varphi(\varrho) \leq \frac{C}{(\varrho_1 - \varrho)^2}\sqrt{\varphi(\varrho_1)} \leq \frac{C}{(\varrho_1 - \varrho)^2}\left(\frac{C}{(\varrho_2 - \varrho_1)^2}\right)^{1/2} M^{1/4}.$$

Iterating this process we get, by induction on n,

$$\varphi(\varrho) \leq \frac{C}{(\varrho_1 - \varrho)^2}\left(\frac{C}{(\varrho_2 - \varrho_1)^2}\right)^{1/2} \cdots \left(\frac{C}{(\varrho_n - \varrho_{n-1})^2}\right)^{1/2^{n-1}} M^{1/2^n} \tag{1.8}$$

for any $0 < \varrho < \varrho_1 < \varrho_2 < \cdots < \varrho_n < R_0$.

For any ϱ with $0 < \varrho < R_0$ we choose

$$\varrho_1 = \frac{\varrho + R_0}{2},\ \varrho_2 = \frac{\varrho_1 + R_0}{2},\ \ldots,\ \varrho_n = \frac{\varrho_{n-1} + R_0}{2}.$$

Then

$$(\varrho_1 - \varrho)^2 = \frac{(R_0 - \varrho)^2}{2^2},\ (\varrho_2 - \varrho_1)^2 = \frac{(R_0 - \varrho)^2}{2^4},\ \ldots,\ (\varrho_n - \varrho_{n-1})^2 = \frac{(R_0 - \varrho)^2}{2^{2n}},$$

whence, by (1.8),

$$\begin{aligned}\varphi(\varrho) &\leq \frac{4C}{(R_0 - \varrho)^2}\left(\frac{4^2 C}{(R_0 - \varrho)^2}\right)^{1/2} \cdots \left(\frac{4^n C}{(R_0 - \varrho)^2}\right)^{1/2^{n-1}} M^{1/2^n} \\ &= 4^{1+2(1/2)+\cdots+n(1/2^{n-1})}\left(\frac{C}{(R_0 - \varrho)^2}\right)^{1+1/2+\cdots+1/2^{n-1}} M^{1/2^n}.\end{aligned}$$

For $n \to +\infty$ we get

$$\begin{aligned}\varphi(\varrho) &\leq 4^{1+2(1/2)+3(1/2^2)+\cdots} \left(\frac{C}{(R_0-\varrho)^2}\right)^{1+1/2+1/2^2+\cdots} \\ &= 4^{(1-1/2)^{-2}} \left(\frac{C}{(R_0-\varrho)^2}\right)^{(1-1/2)^{-1}} = 4^4 \left(\frac{C}{(R_0-\varrho)^2}\right)^2. \end{aligned}$$ □

We now proceed to a proof of Schottky's theorem, which implies a strong constraint for a function regular in a disc and omitting two distinct values.

Theorem 1.2 (Schottky) *Let the function $f(z)$ be regular and satisfy $f(z) \neq 0$ and $f(z) \neq 1$ for all z in the disc $|z| \leq R_0$. Then, for any z, R such that $|z| \leq R < R_0$,*

$$|f(z)| \leq \exp \frac{K R_0^4}{(R_0 - R)^4},$$

where $K > 0$ is a constant depending only on $f(0)$.

Proof Let

$$g_1(z) = \log f(z), \qquad g_2(z) = \log(1 - f(z)),$$

where the logarithms take their principal values at $z = 0$, i.e.,

$$-\pi < \operatorname{Im} g_1(0) = \arg f(0) \leq \pi, \quad -\pi < \operatorname{Im} g_2(0) = \arg(1 - f(0)) \leq \pi, \quad (1.9)$$

and are defined by continuity in the disc $|z| \leq R_0$. By the assumption $f(z) \neq 0, 1$, in the disc $|z| \leq R_0$ the functions $g_1(z)$ and $g_2(z)$ have no branch points, and hence are regular and $\neq 2k\pi i$ ($k \in \mathbb{Z}$). For any r such that $0 < r \leq R_0$, let

$$\begin{aligned} M_1(r) &= \max_{|z|=r} |g_1(z)|, \quad M_2(r) = \max_{|z|=r} |g_2(z)|, \\ M(r) &= \max\{M_1(r), M_2(r)\}, \end{aligned}$$

and

$$G_1(r) = -\min_{|z|=r} \operatorname{Re} g_1(z) = \max_{|z|=r} \log \frac{1}{|f(z)|}.$$

We apply Borel–Carathéodory's inequality (1.1) to the function $-g_1(z)$. We get

$$M_1(\varrho) \leq \frac{2\varrho}{r - \varrho} G_1(r) + \frac{r + \varrho}{r - \varrho} |g_1(0)| \qquad (1.10)$$

for any ϱ, r such that

$$0 < \varrho < r < R < R_0.$$

If

$$G_1(r) > 1, \tag{1.11}$$

let z_0 be such that $|z_0| = r$ and

$$G_1(r) = \log \frac{1}{|f(z_0)|}. \tag{1.12}$$

Then

$$|f(z_0)| = e^{-G_1(r)} < e^{-1} < \frac{1}{2}. \tag{1.13}$$

Thus there exist a value of the logarithm, and hence an integer k, such that

$$-\sum_{n=1}^{\infty} \frac{f(z_0)^n}{n} = \log(1 - f(z_0)) = g_2(z_0) - 2k\pi i. \tag{1.14}$$

By (1.13) we have $1 - |f(z_0)| > 1/2$, whence

$$\sum_{n=1}^{\infty} |f(z_0)|^n = \frac{|f(z_0)|}{1 - |f(z_0)|} < 2\,|f(z_0)| < 1.$$

It follows that, by (1.14),

$$|g_2(z_0) - 2k\pi i| < \sum_{n=1}^{\infty} |f(z_0)|^n < 2\,|f(z_0)| < 1, \tag{1.15}$$

whence

$$2|k|\pi - |g_2(z_0)| \leq |g_2(z_0) - 2k\pi i| < 1.$$

Therefore

$$2|k|\pi < 1 + |g_2(z_0)| \leq 1 + M_2(r) \leq 1 + M_2(R). \tag{1.16}$$

Let

$$h(z) = \log(g_2(z) - 2k\pi i), \tag{1.17}$$

where the logarithm takes its principal value at $z = 0$, whence

$$|\operatorname{Im} h(0)| \leq \pi. \tag{1.18}$$

Since $g_2(z) \neq 2k\pi i$ for $|z| \leq R_0$, $h(z)$ is regular in the disc $|z| \leq R_0$. We apply Borel–Carathéodory's inequality (1.1) to $h(z)$. We obtain

$$\max_{|z|=r} |h(z)| \leq \frac{2r}{R-r} \max_{|z|=R} \log|g_2(z) - 2k\pi i| + \frac{R+r}{R-r}|h(0)|. \tag{1.19}$$

By (1.12), (1.15) and (1.17),

$$\begin{aligned} G_1(r) - \log 2 &= \log \frac{1}{2|f(z_0)|} < \log \frac{1}{|g_2(z_0) - 2k\pi i|} \\ &\leq \left| \log \frac{1}{g_2(z_0) - 2k\pi i} \right| = |-h(z_0)| \leq \max_{|z|=r} |h(z)|. \end{aligned} \tag{1.20}$$

By (1.16),

$$\begin{aligned} \max_{|z|=R} \log|g_2(z) - 2k\pi i| &\leq \max_{|z|=R} \log(|g_2(z)| + 2|k|\pi) \\ &= \log(M_2(R) + 2|k|\pi) < \log(2M_2(R) + 1). \end{aligned} \tag{1.21}$$

Also, by (1.17) and (1.18),

$$|h(0)| \leq |\operatorname{Re} h(0)| + |\operatorname{Im} h(0)| \leq \big|\log|g_2(0) - 2k\pi i|\big| + \pi. \tag{1.22}$$

If $k = 0$ we have

$$|h(0)| \leq \big|\log|g_2(0)|\big| + \pi. \tag{1.23}$$

If $k \neq 0$ we have, by (1.9),

$$\begin{aligned} |g_2(0) - 2k\pi i| &\geq |\operatorname{Im}(g_2(0) - 2k\pi i)| = |\operatorname{Im} g_2(0) - 2k\pi| \\ &\geq 2|k|\pi - |\operatorname{Im} g_2(0)| \geq 2\pi - \pi = \pi > 1, \end{aligned}$$

whence $\log|g_2(0) - 2k\pi i| > 0$. From (1.16), (1.22) and the maximum principle we get, for $k \neq 0$,

$$\begin{aligned} |h(0)| &\leq \log|g_2(0) - 2k\pi i| + \pi \leq \log(|g_2(0)| + 2|k|\pi) + \pi \\ &\leq \log(|g_2(0)| + 1 + M_2(R)) + \pi \leq \log(2M_2(R) + 1) + \pi. \end{aligned} \tag{1.24}$$

Combining (1.23) and (1.24) we obtain, for any value of the integer k,

$$|h(0)| \leq \log(2M_2(R) + 1) + \big|\log|g_2(0)|\big| + \pi,$$

whence, by (1.19) and (1.21),

$$\max_{|z|=r} |h(z)| \le \frac{2R}{R-r}\left(2\log(2M_2(R)+1) + \big|\log|g_2(0)|\big| + \pi\right). \tag{1.25}$$

By (1.20) and (1.25),

$$\begin{aligned} G_1(r) &\le \frac{2R}{R-r}\left(2\log(2M_2(R)+1) + \big|\log|g_2(0)|\big| + \pi\right) + \log 2 \\ &< \frac{2R}{R-r}\left(2\log(2M_2(R)+1) + \big|\log|g_2(0)|\big| + 2\pi\right) \\ &< \frac{4R}{R-r}\left(\log(2M_2(R)+1) + \big|\log|g_2(0)|\big| + \pi\right). \end{aligned}$$

The last inequality is proved under the assumption (1.11), but is obviously true if $G_1(r) \le 1$. Therefore, by (1.10),

$$\begin{aligned} M_1(\varrho) &\le \frac{2r}{r-\varrho}\left(\frac{4R}{R-r}\left(\log(2M_2(R)+1) + \big|\log|g_2(0)|\big| + \pi\right) + |g_1(0)|\right) \\ &< \frac{8Rr}{(R-r)(r-\varrho)}\left(\log(2M_2(R)+1) + |g_1(0)| + \big|\log|g_2(0)|\big| + \pi\right). \end{aligned} \tag{1.26}$$

We have

$$\begin{aligned} \log(2M_2(R)+1) &\le \begin{cases} \log(3M_2(R)) = \log M_2(R) + \log 3, & \text{if } M_2(R) \ge 1 \\ \log 3, \quad \text{if } M_2(R) < 1 \end{cases} \\ &= \log^+ M_2(R) + \log 3, \end{aligned}$$

where $\log^+ x = \max\{\log x,\ 0\}$ for $x > 0$. Clearly $\log^+ x < \sqrt{x}$. Therefore

$$\log(2M_2(R)+1) < \sqrt{M_2(R)} + \log 3.$$

From (1.26) we get

$$M_1(\varrho) < \frac{8Rr}{(R-r)(r-\varrho)}\left(\sqrt{M_2(R)} + K_1\right) \le \frac{8Rr}{(R-r)(r-\varrho)}\left(\sqrt{M(R)} + K_1\right),$$

with

$$K_1 = |g_1(0)| + \big|\log|g_2(0)|\big| + \pi + \log 3.$$

We can repeat the above argument using the function $1 - f(z)$ in place of $f(z)$, i.e., interchanging $g_1(z)$ and $g_2(z)$. Thus

$$M_2(\varrho) < \frac{8Rr}{(R-r)(r-\varrho)}\left(\sqrt{M_1(R)} + K_2\right) \le \frac{8Rr}{(R-r)(r-\varrho)}\left(\sqrt{M(R)} + K_2\right),$$

with

$$K_2 = |g_2(0)| + \left|\log|g_1(0)|\right| + \pi + \log 3.$$

Hence

$$M(\varrho) < \frac{8Rr}{(R-r)(r-\varrho)}\left(\sqrt{M(R)} + \widetilde{K}\right), \tag{1.27}$$

where

$$\widetilde{K} = \max\{K_1, K_2\}.$$

Let $K_* = \max\{|g_1(0)|, |g_2(0)|\}$, and $K_0 = 1 + \widetilde{K}/\sqrt{K_*}$. Then, by the maximum principle,

$$\frac{\widetilde{K}^2}{(K_0-1)^2} = K_* \le M(R).$$

Therefore

$$\widetilde{K} \le (K_0 - 1)\sqrt{M(R)},$$

whence

$$\sqrt{M(R)} + \widetilde{K} \le K_0\sqrt{M(R)}.$$

From (1.27) we obtain

$$M(\varrho) < \frac{8K_0Rr}{(R-r)(r-\varrho)}\sqrt{M(R)} < \frac{8K_0R_0^2}{(R-r)(r-\varrho)}\sqrt{M(R)}, \tag{1.28}$$

where K_0 is a constant depending only on $|g_1(0)|$ and $|g_2(0)|$, i.e., on $f(0)$.

With the choice $r = (R+\varrho)/2$ we have $R - r = r - \varrho = (R-\varrho)/2$, whence, by (1.28),

$$M(\varrho) < \frac{32K_0R_0^2}{(R-\varrho)^2}\sqrt{M(R)}.$$

Since $M(\varrho) < M(R_0)$, Lemma 1.1 yields

$$M(\sigma) \leq \frac{2^{18} K_0^2 R_0^4}{(R_0 - \sigma)^4} \leq \frac{2^{18} K_0^2 R_0^4}{(R_0 - R)^4}$$

for any σ satisfying $0 < \sigma \leq R < R_0$. Thus, for $|z| = \sigma$,

$$\log |f(z)| \leq |g_1(z)| \leq M_1(\sigma) \leq M(\sigma) \leq \frac{2^{18} K_0^2 R_0^4}{(R_0 - R)^4},$$

whence

$$|f(z)| \leq \exp \frac{2^{18} K_0^2 R_0^4}{(R_0 - R)^4} = \exp \frac{K R_0^4}{(R_0 - R)^4}.$$ □

1.3 Picard's First Theorem

Theorem 1.3 (Picard's first theorem) *Let the function*

$$f : \mathbb{C} \longrightarrow \mathbb{C}$$

be entire (i.e., regular in $\mathbb{C}$*) and not constant. Then one of the following two cases occurs: either* $f(\mathbb{C}) = \mathbb{C}$*, or* $f(\mathbb{C}) = \mathbb{C} \setminus \{a\}$ *for an exceptional* $a \in \mathbb{C}$ *depending on* f.

Proof Let $f(z)$ be entire, and let $a, b \in \mathbb{C}$, $a \neq b$, be such that $f(z) \neq a$ and $f(z) \neq b$ for all $z \in \mathbb{C}$. We prove that $f(z)$ is constant. Let

$$g(z) = \frac{f(z) - a}{b - a}.$$

Plainly $g(z)$ is entire and satisfies $g(z) \neq 0$ and $g(z) \neq 1$ for all $z \in \mathbb{C}$. By Theorem 1.2 there exists a constant $K > 0$ such that

$$|g(z)| \leq \exp \frac{K R_0^4}{(R_0 - R)^4}$$

for all $0 < R < R_0$ and all z in the disc $|z| \leq R$. Choosing $R_0 = 2R$ and taking R arbitrarily large, we get $|g(z)| \leq \exp(2^4 K)$ for all $z \in \mathbb{C}$. By Liouville's theorem $g(z)$ is constant, and so is $f(z)$. □

Since $|e^z| = e^{\operatorname{Re} z} > 0$ for all $z \in \mathbb{C}$, for the entire function e^z the exceptional a exists and is zero. For other elementary entire functions the exceptional a does not exist: instances of these are the polynomials of degrees ≥ 1 (by the fundamental theorem of algebra), or the circular functions $\sin z$ and $\cos z$. For any $a \in \mathbb{C}$, the equation

$$\sin z = \frac{e^{iz} - e^{-iz}}{2i} = a$$

is equivalent to $e^{2iz} - 2ia\, e^{iz} - 1 = 0$, whence $e^{iz} = ia \pm \sqrt{1-a^2}$, and therefore has the solutions

$$z = \arcsin a = -i \log\left(ia \pm \sqrt{1-a^2}\right)$$

because $ia \neq \mp\sqrt{1-a^2}$ since $-a^2 \neq 1 - a^2$. Similarly, for any $a \in \mathbb{C}$ the equation $\cos z = a$ is solvable.

1.4 Picard's Second Theorem

Here and in what follows, we say that a function $f(z)$ has an *isolated essential singularity* at $z_0 \in \mathbb{C}$ if, for some $\varrho > 0$, in the punctured disc

$$\mathcal{P}_{z_0,\varrho} := \{z \in \mathbb{C} \mid 0 < |z - z_0| < \varrho\} \tag{1.29}$$

the function $f(z)$ is regular and has a Laurent series expansion

$$f(z) = \sum_{n=-\infty}^{+\infty} a_n (z - z_0)^n \tag{1.30}$$

with $a_n \neq 0$ for infinitely many $n < 0$. We require the following

Lemma 1.2 (Weierstrass) *Let $z_0 \in \mathbb{C}$ be an isolated essential singularity for $f(z)$. For any $\varepsilon > 0$, the set $f(\mathcal{P}_{z_0,\varepsilon})$ of the values taken by $f(z)$ in the punctured disc $\mathcal{P}_{z_0,\varepsilon} = \{z \in \mathbb{C} \mid 0 < |z - z_0| < \varepsilon\}$ is dense in $\mathbb{C}$.*

Proof By contradiction. Assume there exist $\varrho > 0$, $\eta > 0$ and $a \in \mathbb{C}$ such that $f(z)$ is regular in the punctured disc (1.29) and satisfies $|f(z) - a| \geq \eta$ for all $z \in \mathcal{P}_{z_0,\varrho}$. Then in $\mathcal{P}_{z_0,\varrho}$ the function $g(z) := (f(z) - a)^{-1}$ is regular and satisfies

$$|g(z)| = \frac{1}{|f(z) - a|} \leq \frac{1}{\eta}. \tag{1.31}$$

Denote by

$$g(z) = \sum_{n=-\infty}^{+\infty} b_n (z - z_0)^n \tag{1.32}$$

the Laurent series expansion of $g(z)$ in $\mathcal{P}_{z_0,\varrho}$. Then, for any $n \in \mathbb{Z}$ and any δ such that $0 < \delta < \varrho$,

$$b_n = \frac{1}{2\pi i} \oint_{|z-z_0|=\delta} \frac{g(z)}{(z-z_0)^{n+1}}\, dz.$$

For $n = -(k+1)$ with $k \geq 0$ we get, by (1.31),

$$|b_n| = \left|b_{-(k+1)}\right| = \left|\frac{1}{2\pi i} \oint_{|z-z_0|=\delta} g(z)(z-z_0)^k dz\right|$$
$$\leq \frac{\delta^k}{2\pi\eta} \cdot 2\pi\delta = \frac{\delta^{k+1}}{\eta} \to 0 \quad (\delta \to 0),$$

whence $b_{-1} = b_{-2} = b_{-3} = \ldots = 0$, and (1.32) is a Taylor series. Thus $g(z)$ is regular at z_0, whence $f(z) = a + 1/g(z)$ either is regular or has a pole at z_0, contradicting the assumption that z_0 is an essential singularity for $f(z)$. □

Theorem 1.4 (Picard's second theorem) *Let $z_0 \in \mathbb{C}$ be an isolated essential singularity for $f(z)$. For any $\varepsilon > 0$, the set $f(\mathcal{P}_{z_0,\varepsilon})$ of the values taken by $f(z)$ in the punctured disc $\mathcal{P}_{z_0,\varepsilon} = \{z \in \mathbb{C} \mid 0 < |z - z_0| < \varepsilon\}$ is either $\mathbb{C}$, or $\mathbb{C} \setminus \{a\}$ for an exceptional $a \in \mathbb{C}$ depending on f and on the essential singularity z_0, but independent of ε.*

Proof By contradiction. Assume there exist $\varrho > 0$ and $a, b \in \mathbb{C}$, with $a \neq b$, such that $f(z)$ is regular in the punctured disc (1.29) and satisfies $f(z) \neq a$ and $f(z) \neq b$ for all $z \in \mathcal{P}_{z_0,\varrho}$. We claim that there exists a sequence of positive numbers r_ν, with $\varrho > r_1 > r_2 > r_3 > \cdots \to 0$, such that

$$|f(z)| \leq C \text{ for } |z - z_0| = r_\nu \quad (\nu = 1, 2, 3, \ldots), \tag{1.33}$$

where C is a constant independent of ν. For convenience, by a translation we may assume $z_0 = 0$. Then, by replacing $f(z)$ with

$$\frac{f(\varrho z) - a}{b - a},$$

we may also assume $\varrho = 1$, $a = 0$ and $b = 1$. By Lemma 1.2 there exists a sequence $z_1, z_2, z_3, \ldots \in \mathcal{P}_{0,1}$ such that $|z_1| > |z_2| > |z_3| > \cdots \to 0$ and

$$|f(z_\nu) - c_1| < c_2 \quad (\nu = 1, 2, 3, \ldots), \tag{1.34}$$

where c_1 and c_2 are any fixed real numbers satisfying $c_1 > 2$ and $0 < c_2 < c_1 - 2$ (for instance, we may take $c_1 = 3$ and $c_2 = \frac{1}{2}$).

Let $r_\nu = |z_\nu|$ $(\nu = 1, 2, 3, \dots)$, and let $w = \log z$ for $z \in \mathcal{P}_{0,1}$. Then $z = e^w$ is a one-to-one mapping of the half strip

$$\mathcal{S} := \{w = u + iv \mid u < 0,\ -\pi < v \le \pi\}$$

onto $\mathcal{P}_{0,1}$, and in the half-plane $u < 0$ the function

$$F(w) := f(e^w)$$

is regular and periodic with period $2\pi i$.

Let $w_\nu = \log z_\nu$ $(\nu = 1, 2, 3, \dots)$ with the principal value of the logarithm, whence $w_\nu = u_\nu + iv_\nu \in \mathcal{S}$ and $u_\nu = \log|z_\nu| \to -\infty$. Thus for any sufficiently large ν we have $u_\nu < -4\pi$. For such ν, the function

$$\widetilde{f}_\nu(t) := F(w_\nu + t) = f(e^{w_\nu + t})$$

is regular in the disc $|t| \le 4\pi$ and satisfies $\widetilde{f}_\nu(t) \ne 0, 1$. By Theorem 1.2 with $R = 2\pi$ and $R_0 = 4\pi$ we get $\left|\widetilde{f}_\nu(t)\right| \le \exp(2^4 K_\nu)$ for $|t| \le 2\pi$, where the constant K_ν depends only on $\widetilde{f}_\nu(0) = f(z_\nu)$. Moreover, with the notation used in the proof of Theorem 1.2, we may take $K_\nu = 2^{18} K_0^2$ with $K_0 = 1 + \widetilde{K}/\sqrt{K_*}$, so that an upper bound for K_ν with an absolute constant is obtained from an upper bound for

$$\widetilde{K} = \max\{|g_1(0)| + \big|\log|g_2(0)|\big|,\ |g_2(0)| + \big|\log|g_1(0)|\big|\} + \pi + \log 3$$

and a lower bound for

$$K_* = \max\{|g_1(0)|,\ |g_2(0)|\},$$

and hence from upper and lower bounds with positive absolute constants for

$$|g_1(0)| = \left|\log \widetilde{f}_\nu(0)\right| = \sqrt{\log^2|f(z_\nu)| + \arg^2 f(z_\nu)}$$

and

$$|g_2(0)| = \left|\log(1 - \widetilde{f}_\nu(0))\right| = \sqrt{\log^2|1 - f(z_\nu)| + \arg^2(1 - f(z_\nu))},$$

where, in accordance with (1.9), arg denotes the principal argument. By virtue of (1.34), K_ν does not exceed an absolute constant. Therefore we get $|F(w)| \le C$, with an absolute constant C, in the disc $|w - w_\nu| \le 2\pi$, and *a fortiori* for $w = u + iv$ satisfying $u = u_\nu = \operatorname{Re} w_\nu = \log|z_\nu|$ and $|v| \le \pi$. It follows that $|f(z)| \le C$ for $|z| = |z_\nu| = r_\nu$, i.e., (1.33).

By the same method used in the proof of Lemma 1.2, (1.33) easily implies that $f(z)$ is regular at z_0. For, denoting by (1.30) the Laurent expansion of $f(z)$ in $\mathcal{P}_{z_0,\varrho}$, we have

$$a_n = \frac{1}{2\pi i} \oint\limits_{|z-z_0|=r_\nu} \frac{f(z)}{(z-z_0)^{n+1}}\,\mathrm{d}z \qquad (n \in \mathbb{Z};\ \nu = 1, 2, 3, \dots).$$

Then, for $n = -(k+1)$ with $k \geq 0$, one gets by (1.33)

$$\begin{aligned} |a_n| = \left|a_{-(k+1)}\right| &= \left| \frac{1}{2\pi i} \oint\limits_{|z-z_0|=r_\nu} f(z)(z-z_0)^k\,\mathrm{d}z \right| \\ &\leq \frac{Cr_\nu^k}{2\pi} \cdot 2\pi r_\nu = Cr_\nu^{k+1} \to 0 \qquad (\nu \to \infty) \end{aligned}$$

whence $a_{-1} = a_{-2} = a_{-3} = \dots = 0$, and (1.30) is a Taylor series, i.e., $f(z)$ is regular at z_0. This contradicts the assumption that z_0 is an essential singularity for $f(z)$, and therefore completes the proof of Picard's second theorem. □

Since the ε in the statement of Theorem 1.4 is arbitrarily small, it follows that for any $b \in \mathbb{C}$, or for any $b \in \mathbb{C} \setminus \{a\}$ if the exceptional a exists, there are infinitely many $z \in \mathcal{P}_{z_0,\varepsilon}$ such that $f(z) = b$.

Chapter 2
The Weierstrass Factorization Theorem

2.1 Mittag-Leffler's Series

A function $f(z)$ is *meromorphic* in an open set $A \subset \mathbb{C}$ if it is regular in A except for a finite or infinite sequence $z_1, z_2, \ldots \in A$ of poles of $f(z)$ (of any multiplicities). For short, in what follows a function meromorphic in the whole $\mathbb{C}$ will simply be called meromorphic.

If $f(z)$ is meromorphic with only finitely many poles $z_1, z_2, \ldots, z_N$ of respective multiplicities $\mu_1, \mu_2, \ldots, \mu_N$, for any $n = 1, 2, \ldots, N$ let the Laurent series expansion of $f(z)$ in the neighbourhood of z_n be denoted by

$$f(z) = \sum_{k=0}^{\infty} a_k^{(n)} (z - z_n)^k + \sum_{k=1}^{\mu_n} \frac{b_k^{(n)}}{(z - z_n)^k}.$$

Then

$$G(z) := f(z) - \sum_{n=1}^{N} \sum_{k=1}^{\mu_n} \frac{b_k^{(n)}}{(z - z_n)^k}$$

is plainly an entire function. Conversely, for any entire function $G(z)$, the function

$$f(z) = G(z) + \sum_{n=1}^{N} \sum_{k=1}^{\mu_n} \frac{b_k^{(n)}}{(z - z_n)^k} \tag{2.1}$$

is meromorphic, with finitely many poles $z_1, \ldots, z_N$ of respective multiplicities $\mu_1, \ldots, \mu_N$.

If a meromorphic function $f(z)$ has infinitely many poles $z_1, z_2, \ldots$, in general the decomposition similar to (2.1) with an entire function $G(z)$ does not hold, because the series

C. Viola, *An Introduction to Special Functions*,
UNITEXT - La Matematica per il 3+2 102, DOI 10.1007/978-3-319-41345-7_2

$$\sum_{n=1}^{\infty}\sum_{k=1}^{\mu_n}\frac{b_k^{(n)}}{(z-z_n)^k} \tag{2.2}$$

is not expected to converge. However, Mittag-Leffler has shown that (2.2) can be suitably modified, so that the series thus obtained is totally convergent in any compact subset of $\mathbb{C}$ not containing the poles $z_1, z_2, \ldots$.

We recall that a series of functions

$$\sum_{n=1}^{\infty} u_n(z)$$

is said to be totally convergent in a set $K \subset \mathbb{C}$, if there exists a sequence of constants $c_n > 0$ such that

$$|u_n(z)| \le c_n \quad \text{for all } n = 1, 2, \ldots \text{ and for all } z \in K,$$

with

$$\sum_{n=1}^{\infty} c_n < +\infty.$$

By Weierstrass' test, if $\sum\limits_{n=1}^{\infty} u_n(z)$ is totally convergent in K, it is absolutely and uniformly convergent in K.

In view of subsequent applications we prove a special case of Mittag-Leffler's theorem, corresponding to the most important case where the infinitely many poles $z_1, z_2, \ldots$ are all simple.

Theorem 2.1 (Mittag-Leffler) *Let $z_1, z_2, \ldots \to \infty$ be a sequence of distinct complex numbers satisfying $0 < |z_1| \le |z_2| \le \ldots$. Let $m_1, m_2, \ldots$ be any sequence of non-zero complex numbers. Then there exists a (not unique) sequence $p_1, p_2, \ldots$ of non-negative integers, depending only on the sequences (z_n) and (m_n), such that the series*

$$f(z) := \sum_{n=1}^{\infty}\left(\frac{z}{z_n}\right)^{p_n}\frac{m_n}{z-z_n} \tag{2.3}$$

is totally convergent, and hence absolutely and uniformly convergent, in any compact set $K \subset \mathbb{C} \setminus \{z_1, z_2, \ldots\}$. Thus the function $f(z)$ is meromorphic, with simple poles $z_1, z_2, \ldots$ having respective residues $m_1, m_2, \ldots$.

Proof We choose a sequence of real numbers $0 < r_1 \leq r_2 \leq \ldots \to +\infty$ satisfying $r_n < |z_n|$ $(n = 1, 2, \ldots)$. In the disc $|z| \leq r_n$ we have

$$\left|\frac{m_n}{z - z_n}\right| \leq \frac{|m_n|}{|z_n| - |z|} \leq \frac{|m_n|}{|z_n| - r_n} \tag{2.4}$$

and

$$\left|\frac{z}{z_n}\right| \leq \frac{r_n}{|z_n|} < 1.$$

Let $\varepsilon_1 + \varepsilon_2 + \ldots$ be any convergent series of positive constants ε_n. For every n, let p_n be any non-negative integer such that

$$\left(\frac{r_n}{|z_n|}\right)^{p_n} < \frac{\varepsilon_n}{|m_n|}(|z_n| - r_n). \tag{2.5}$$

Since $r_n/|z_n| < 1$, (2.5) is satisfied for any sufficiently large p_n. Then in the disc $|z| \leq r_n$ we get by (2.4) and (2.5)

$$\left|\left(\frac{z}{z_n}\right)^{p_n} \frac{m_n}{z - z_n}\right| \leq \left(\frac{r_n}{|z_n|}\right)^{p_n} \frac{|m_n|}{|z_n| - r_n} < \varepsilon_n. \tag{2.6}$$

Take any compact set $K \subset \mathbb{C} \setminus \{z_1, z_2, \ldots\}$, and choose an integer N such that K is contained in the disc $|z| \leq r_N$. Let

$$M_n = \max_{z \in K} \left|\left(\frac{z}{z_n}\right)^{p_n} \frac{m_n}{z - z_n}\right|.$$

Then, by (2.6), for any $z \in K$ we have

$$\left|\left(\frac{z}{z_n}\right)^{p_n} \frac{m_n}{z - z_n}\right| \leq \begin{cases} M_n & \text{if } n < N \\ \varepsilon_n & \text{if } n \geq N. \end{cases}$$

Since the series of constants $M_1 + \ldots + M_{N-1} + \varepsilon_N + \varepsilon_{N+1} + \ldots$ converges, (2.3) is totally convergent in K. □

Remark 2.1 Fix any $z \in \mathbb{C}$, and let n be such that $|z| < |z_n|$. Then $2|z_n| > |z| + |z_n|$, whence

$$\left|\frac{z}{z_n}\right| \leq \frac{2|z|}{|z| + |z_n|} \leq \frac{2|z|}{|z - z_n|},$$

and

$$|m_n| \left|\frac{z}{z_n}\right|^{p_n+1} \leq 2|z| \left|\left(\frac{z}{z_n}\right)^{p_n} \frac{m_n}{z - z_n}\right|.$$

It follows that the sequence (p_n) in Theorem 2.1 is such that

$$\sum_{n=1}^{\infty} |m_n| \left|\frac{z}{z_n}\right|^{p_n+1} < +\infty \quad \text{for every } z \in \mathbb{C}. \tag{2.7}$$

Conversely, any sequence (p_n) of non-negative integers satisfying (2.7) is such that the series (2.3) is totally convergent in any compact set $K \subset \mathbb{C} \setminus \{z_1, z_2, \dots\}$. For, if $z_* \in \mathbb{C}$ is such that $|z_*| > \max_{z \in K} |z|$, for any n satisfying $|z_n| > |z_*| + 1$ we have

$$\frac{|z_n|}{|z_n| - |z_*|} = \frac{|z_*|}{|z_n| - |z_*|} + 1 < |z_*| + 1,$$

whence

$$\frac{1}{|z_n| - |z_*|} < \frac{|z_*| + 1}{|z_n|} = \left(1 + \frac{1}{|z_*|}\right)\left|\frac{z_*}{z_n}\right|.$$

Thus, for any $z \in K$,

$$\left|\left(\frac{z}{z_n}\right)^{p_n} \frac{m_n}{z - z_n}\right| \le \left|\frac{z_*}{z_n}\right|^{p_n} \frac{|m_n|}{|z_n| - |z_*|} \le \left(1 + \frac{1}{|z_*|}\right) |m_n| \left|\frac{z_*}{z_n}\right|^{p_n+1}.$$

By (2.7), the series (2.3) is totally convergent in K.

Remark 2.2 If $g(z)$ is any meromorphic function with infinitely many poles $z_1, z_2, \dots$, all simple and with respective residues $m_1, m_2, \dots$, then for any sequence (p_n) of non-negative integers satisfying (2.7) the function $G(z) := g(z) - f(z)$, where $f(z)$ is given by (2.3), is plainly entire. Thus $g(z)$ has the Mittag-Leffler series expansion

$$g(z) = G(z) + \sum_{n=1}^{\infty} \left(\frac{z}{z_n}\right)^{p_n} \frac{m_n}{z - z_n},$$

where $G(z)$ is an entire function.

Remark 2.3 The function

$$f(z) = \sum_{n=1}^{\infty} \left(\frac{z}{z_n}\right)^{p_n} \frac{m_n}{z - z_n}$$

in (2.3) has the Taylor series expansion around $z = 0$ given by

$$f(z) = -\sum_{k=0}^{\infty} \Bigg(\sum_{\substack{n \\ p_n \le k}} m_n \, z_n^{-(k+1)} \Bigg) z^k \tag{2.8}$$

with radius of convergence $|z_1|$. For, in the disc $|z| \le r_1 < |z_1|$ we have

$$f(z) = -\sum_{n=1}^{\infty} m_n \frac{z^{p_n}}{z_n^{p_n+1}} \frac{1}{1 - z/z_n} = -\sum_{n=1}^{\infty} m_n \sum_{k=p_n}^{\infty} \frac{z^k}{z_n^{k+1}}. \tag{2.9}$$

Since, by (2.5),

$$\sum_{n=1}^{\infty}\sum_{k=p_n}^{\infty} |m_n| \frac{|z|^k}{|z_n|^{k+1}} = \sum_{n=1}^{\infty} |m_n| \frac{|z|^{p_n}}{|z_n|^{p_n+1}} \frac{1}{1 - |z|/|z_n|} = \sum_{n=1}^{\infty} \left|\frac{z}{z_n}\right|^{p_n} \frac{|m_n|}{|z_n| - |z|}$$
$$\le \sum_{n=1}^{\infty} \left(\frac{r_n}{|z_n|}\right)^{p_n} \frac{|m_n|}{|z_n| - r_n} \le \sum_{n=1}^{\infty} \varepsilon_n < +\infty,$$

we may interchange the sums on the right-hand side of (2.9). Therefore

$$f(z) = -\sum_{k=0}^{\infty} z^k \sum_{\substack{n \\ p_n \le k}} m_n\, z_n^{-(k+1)},$$

and the radius of convergence of this Taylor series is the distance from the origin to the closest singular point of $f(z)$, i.e., to the pole z_1.

2.2 Infinite Products

Let $a_n \in \mathbb{C}$, $a_n \neq 0$ $(n = 1, 2, \dots)$. The infinite product $\prod_n a_n$ is defined by

$$\prod_{n=1}^{\infty} a_n = \lim_{n\to\infty} \prod_{k=1}^{n} a_k = A. \tag{2.10}$$

The infinite product (2.10) converges if the limit A exists and $A \neq 0, \infty$. If $A = 0$ or $A = \infty$, the infinite product is said to be divergent to zero or, respectively, to infinity.

If (2.10) converges, then

$$\lim_{n\to\infty} a_n = \lim_{n\to\infty} \left(\prod_{k=1}^{n} a_k \Big/ \prod_{k=1}^{n-1} a_k \right) = A/A = 1.$$

Let $a_n(z) \neq 0$ $(n = 1, 2, \dots)$ be a sequence of functions defined for z in a set $K \subset \mathbb{C}$. The infinite product $\prod_n a_n(z)$ is uniformly convergent to a function $f(z) \neq 0, \infty$ in K if, for $n \to \infty$, the sequence of partial products

$$f_n(z) = \prod_{k=1}^{n} a_k(z)$$

converges uniformly to $f(z)$ in K, i.e., if for any $\varepsilon > 0$ there exists n_0 such that

$$|f(z) - f_n(z)| < \varepsilon$$

for all $n > n_0$ and for all $z \in K$.

Lemma 2.1 *Let the functions $u_n(z)$ $(n = 1, 2, \dots)$ be regular in a compact set $K \subset \mathbb{C}$, and let the series*

$$\sum_{n=1}^{\infty} u_n(z)$$

be totally convergent in K. Then the infinite product

$$\prod_{n=1}^{\infty} \exp\left(u_n(z)\right) = \exp\left(\sum_{n=1}^{\infty} u_n(z)\right)$$

is uniformly convergent in K.

Proof For any $z_1, z_2 \in \mathbb{C}$ we have

$$\begin{aligned} e^{z_1} - e^{z_2} &= (z_1 - z_2) + \frac{1}{2!}(z_1^2 - z_2^2) + \frac{1}{3!}(z_1^3 - z_2^3) + \dots \\ &= (z_1 - z_2)\Big(1 + \frac{1}{2!}(z_1 + z_2) + \frac{1}{3!}(z_1^2 + z_1 z_2 + z_2^2) + \dots\Big), \end{aligned}$$

whence

$$\begin{aligned} |e^{z_1} - e^{z_2}| &\le |z_1 - z_2|\Big(1 + \frac{1}{2!}(|z_1| + |z_2|) + \frac{1}{3!}(|z_1| + |z_2|)^2 + \dots\Big) \\ &\le |z_1 - z_2|\, e^{|z_1|+|z_2|}. \end{aligned}$$

Therefore

$$\begin{aligned} &\left|\exp\Big(\sum_{n=1}^{\infty} u_n(z)\Big) - \exp\Big(\sum_{n=1}^{N} u_n(z)\Big)\right| \qquad (2.11) \\ &\quad \le \left|\sum_{n=1}^{\infty} u_n(z) - \sum_{n=1}^{N} u_n(z)\right| \exp\Big(\sum_{n=1}^{\infty} |u_n(z)| + \sum_{n=1}^{N} |u_n(z)|\Big) \\ &\quad \le \left|\sum_{n=1}^{\infty} u_n(z) - \sum_{n=1}^{N} u_n(z)\right| \exp\Big(2\sum_{n=1}^{\infty} |u_n(z)|\Big). \end{aligned}$$

Let

$$\sum_{n=1}^{\infty} |u_n(z)| = U(z),$$

let

$$M = \exp\left(2 \max_{z\in K} U(z)\right),$$

and for any $\varepsilon > 0$ let N_0 be such that

$$\left|\sum_{n=1}^{\infty} u_n(z) - \sum_{n=1}^{N} u_n(z)\right| < \frac{\varepsilon}{M}$$

for all $N > N_0$ and all $z \in K$. Then, by (2.11),

$$\left|\exp\left(\sum_{n=1}^{\infty} u_n(z)\right) - \exp\left(\sum_{n=1}^{N} u_n(z)\right)\right| \le M \left|\sum_{n=1}^{\infty} u_n(z) - \sum_{n=1}^{N} u_n(z)\right| < \varepsilon. \quad \square$$

2.3 Weierstrass' Products

Lemma 2.2 *Let $f(z)$ be a meromorphic function. Let $z_1, z_2, \ldots \neq 0$ be the poles of $f(z)$, all simple with respective residues $m_1, m_2, \ldots \in \mathbb{Z}$. Then the function*

$$\varphi(z) := \exp \int_0^z f(t)\,\mathrm{d}t \tag{2.12}$$

is meromorphic. The zeros (resp. poles) of $\varphi(z)$ are the points z_n such that $m_n > 0$ (resp. $m_n < 0$), and the multiplicity of z_n as a zero (resp. pole) of $\varphi(z)$ is m_n (resp. $-m_n$).

Proof First we remark that $\varphi(z)$ is a one-valued function, because if γ and γ' are any two paths of endpoints 0 and z not passing through the poles z_n, by the residue theorem we get

$$\int_\gamma f(t)\,\mathrm{d}t = \int_{\gamma'} f(t)\,\mathrm{d}t + 2\pi i R,$$

where R is the sum of residues of $f(t)$ at the poles between γ and γ', each residue being taken with + or − sign according to the mutual position of γ and γ' around the corresponding pole. Since $m_1, m_2, \ldots \in \mathbb{Z}$, we get $R \in \mathbb{Z}$. Therefore

$$\exp \int_{\gamma} f(t)\,\mathrm{d}t = e^{2\pi i R} \exp \int_{\gamma'} f(t)\,\mathrm{d}t = \exp \int_{\gamma'} f(t)\,\mathrm{d}t.$$

Plainly $\varphi(z)$ is regular and $\neq 0$ in $\mathbb{C} \setminus \{z_1, z_2, \ldots\}$. The function $f_1(z) := f(z) - m_1/(z - z_1)$ is regular in $\mathbb{C} \setminus \{z_2, z_3, \ldots\}$, whence $\exp \int_0^z f_1(t)\,\mathrm{d}t$ is regular and $\neq 0$ in $\mathbb{C} \setminus \{z_2, z_3, \ldots\}$. Thus

$$\begin{aligned}\varphi(z) &= \exp \int_0^z f(t)\,\mathrm{d}t = \exp\left(\int_0^z f_1(t)\,\mathrm{d}t + m_1 \int_0^z \frac{\mathrm{d}t}{t - z_1}\right) \\ &= \exp \int_0^z f_1(t)\,\mathrm{d}t \cdot \exp\left(m_1 \log\left(1 - \frac{z}{z_1}\right)\right) = (z - z_1)^{m_1} \varphi_1(z),\end{aligned}$$

where $\varphi_1(z) = (-z_1)^{-m_1} \exp \int_0^z f_1(t)\,\mathrm{d}t$ is regular and $\neq 0$ in $\mathbb{C} \setminus \{z_2, z_3, \ldots\}$. By the same argument for $z_2, z_3, \ldots$ the lemma follows. □

For any sequence $z_1, z_2, \ldots \in \mathbb{C}$, either finite or satisfying $\lim z_n = \infty$, from Theorem 2.1 and Lemma 2.2 we deduce the existence of a meromorphic function with zeros and poles at $z_1, z_2, \ldots$ with arbitrary multiplicities $m_1, m_2, \ldots$. This yields the following

Corollary 2.1 *Every meromorphic function is the quotient of two entire functions.*

Proof Let $g(z)$ be meromorphic, and let $z_1, z_2, \ldots \neq 0$ be the poles of $g(z)$ with respective multiplicities $m_1, m_2, \ldots$. Let $f(z)$ be the function (2.3) and let $\varphi(z)$ be the function (2.12). By applying Theorem 2.1 and Lemma 2.2, we see that $\varphi(z)$ is entire with zeros $z_1, z_2, \ldots$ of respective multiplicities $m_1, m_2, \ldots$. Then the product $g(z)\varphi(z)$ has no poles, and therefore is an entire function $h(z)$, whence $g(z) = h(z)/\varphi(z)$ with $h(z)$ and $\varphi(z)$ entire functions.

If $g(z)$ has a pole at $z = 0$ of multiplicity m, then $\widetilde{g}(z) := z^m g(z)$ is regular at $z = 0$, whence $\widetilde{g}(z) = h(z)/\varphi(z)$ with entire $h(z)$ and $\varphi(z)$, and $g(z) = h(z)/(z^m \varphi(z))$ with entire $h(z)$ and $z^m \varphi(z)$. □

Theorem 2.2 (Weierstrass) *Let $F(z)$ be meromorphic, and regular and $\neq 0$ at $z = 0$. Let $z_1, z_2, \ldots$ be the zeros and poles of $F(z)$ with respective multiplicities $|m_1|, |m_2|, \ldots$, where $m_n > 0$ if z_n is a zero and $m_n < 0$ if z_n is a pole of $F(z)$. Then there exist integers $p_1, p_2, \ldots \geq 0$ and an entire function $G(z)$ such that*

$$F(z) = e^{G(z)} \prod_n \left(1 - \frac{z}{z_n}\right)^{m_n} \exp\left(m_n \sum_{k=1}^{p_n} \frac{1}{k}\left(\frac{z}{z_n}\right)^k\right), \tag{2.13}$$

where the product converges uniformly in any compact set $K \subset \mathbb{C} \setminus \{z_1, z_2, \dots\}$. In (2.13) *one can take any sequence of integers $p_1, p_2, \dots \geq 0$ satisfying* (2.7).

Proof Let $f(z)$ be the function (2.3) with integer exponents $p_1, p_2, \dots \geq 0$ satisfying (2.7), and let $\varphi(z)$ be the function (2.12). By Theorem 2.1 and Lemma 2.2 $\varphi(z)$ is meromorphic, with zeros z_n of multiplicities m_n if $m_n > 0$, and with poles z_n of multiplicities $|m_n|$ if $m_n < 0$. Thus $F(z)$ and $\varphi(z)$ have the same zeros and poles with the same multiplicities, whence $F(z)/\varphi(z)$ is entire and $\neq 0$. Therefore $\log\big(F(z)/\varphi(z)\big) = G(z)$ is an entire function, and

$$F(z) = e^{G(z)}\varphi(z). \tag{2.14}$$

By uniform convergence of (2.3) on a path of endpoints 0 and z not containing $z_1, z_2, \dots$, term-by-term integration of (2.3) from 0 to z is allowed. Thus from (2.12) we get

$$\begin{aligned}
\varphi(z) &= \exp \int_0^z \sum_n \left(\frac{t}{z_n}\right)^{p_n} \frac{m_n}{t - z_n}\,\mathrm{d}t \qquad (2.15)\\
&= \prod_n \exp \int_0^z \left(\frac{m_n}{t - z_n} + \frac{m_n}{z_n} \frac{(t/z_n)^{p_n} - 1}{t/z_n - 1} \right) \mathrm{d}t \\
&= \prod_n \exp \int_0^z \left(\frac{m_n}{t - z_n} + \frac{m_n}{z_n} \sum_{k=1}^{p_n} \left(\frac{t}{z_n}\right)^{k-1} \right) \mathrm{d}t \\
&= \prod_n \exp \left(\log\left(1 - \frac{z}{z_n}\right)^{m_n} + m_n \sum_{k=1}^{p_n} \frac{1}{k}\left(\frac{z}{z_n}\right)^k \right) \\
&= \prod_n \left(1 - \frac{z}{z_n}\right)^{m_n} \exp\left(m_n \sum_{k=1}^{p_n} \frac{1}{k}\left(\frac{z}{z_n}\right)^k \right).
\end{aligned}$$

Then (2.13) follows from (2.14). Moreover, for any $z \in K$ we can choose the integration path from 0 to z of length bounded by a constant depending only on K, whence, by Theorem 2.1, the series

$$\sum_n \int_0^z \left(\frac{t}{z_n}\right)^{p_n} \frac{m_n}{t - z_n}\,\mathrm{d}t$$

is totally convergent in K. Hence, by Lemma 2.1, the product in (2.13) is uniformly convergent in K. □

If the meromorphic function $F(z)$ has a zero of multiplicity m or a pole of multiplicity $-m$ at $z = 0$, then $z^{-m}F(z)$ is regular and $\neq 0$ at $z = 0$, and therefore can be represented by (2.13). Thus in this case the Weierstrass factorization formula (2.13) becomes

$$F(z) = z^m e^{G(z)} \prod_n \left(1 - \frac{z}{z_n}\right)^{m_n} \exp\left(m_n \sum_{k=1}^{p_n} \frac{1}{k}\left(\frac{z}{z_n}\right)^k\right). \tag{2.16}$$

Corollary 2.2 *The Taylor series expansion with centre $z = 0$ of the entire function $G(z)$ in (2.13) is*

$$G(z) = \log F(0) + \sum_{k=1}^{\infty} \left(\frac{1}{(k-1)!}\left[\frac{\mathrm{d}^{k-1}}{\mathrm{d}z^{k-1}} \frac{F'(z)}{F(z)}\right]_{z=0} + \sum_{\substack{n \\ p_n < k}} m_n z_n^{-k}\right) \frac{z^k}{k}. \tag{2.17}$$

Proof From (2.8), (2.12), (2.13) and (2.15) we get, for $|z| < |z_1|$,

$$\begin{aligned}
G(z) &= \log F(z) - \log \varphi(z) = \log F(z) - \int_0^z f(t)\,\mathrm{d}t \\
&= \log F(z) + \int_0^z \sum_{k=0}^{\infty} \left(\sum_{\substack{n \\ p_n \le k}} m_n z_n^{-(k+1)}\right) t^k\,\mathrm{d}t \\
&= \log F(0) + \sum_{k=1}^{\infty} \frac{z^k}{k!}\left[\frac{\mathrm{d}^k}{\mathrm{d}z^k} \log F(z)\right]_{z=0} + \sum_{k=0}^{\infty} \left(\sum_{\substack{n \\ p_n \le k}} m_n z_n^{-(k+1)}\right) \frac{z^{k+1}}{k+1} \\
&= \log F(0) + \sum_{k=1}^{\infty} \frac{z^k}{k!}\left[\frac{\mathrm{d}^{k-1}}{\mathrm{d}z^{k-1}} \frac{F'(z)}{F(z)}\right]_{z=0} + \sum_{k=1}^{\infty} \left(\sum_{\substack{n \\ p_n < k}} m_n z_n^{-k}\right) \frac{z^k}{k},
\end{aligned}$$

and (2.17) follows. □

By Corollary 2.1, the Weierstrass factorization (2.13) or (2.16) of a meromorphic function $F(z)$ can be reduced to the Weierstrass factorizations of two entire functions $F_1(z)$ and $F_2(z)$ such that $F(z) = F_1(z)/F_2(z)$. Thus, in what follows, we will employ (2.13) or (2.16) only for entire functions $F(z)$, i.e., when $m_n > 0$ ($n = 1, 2, \ldots$) and $m \geq 0$. Moreover, with no loss of generality, in the product on the right-hand sides of (2.13) and (2.16) we can put $m_n = 1$, with the convention that if a zero z_n has multiplicity m_n, the corresponding factor

$$\left(1 - \frac{z}{z_n}\right) \exp \sum_{k=1}^{p_n} \frac{1}{k}\left(\frac{z}{z_n}\right)^k$$

is repeated m_n times in the product. Accordingly, condition (2.7) for the sequence (p_n) becomes

$$\sum_{n=1}^{\infty} \left| \frac{z}{z_n} \right|^{p_n+1} < +\infty \quad \text{for every } z \in \mathbb{C}, \tag{2.18}$$

where if z_n is a zero of multiplicity m_n, the term $|z/z_n|^{p_n+1}$ is repeated m_n times in the series.

Chapter 3
Entire Functions of Finite Order

From now on, we shall use Vinogradov's asymptotic symbol $\ll$ which is defined as follows. Given two functions $f(z)$ and $g(z)$, with $g(z) > 0$, we write

$$f(z) \ll g(z) \quad (z \to z_0), \tag{3.1}$$

where z_0 can be either finite or infinity, to mean that there exists a constant $C > 0$, independent of z, such that $|f(z)| \leq Cg(z)$ for all z in a suitable neighbourhood of z_0. Thus (3.1) is synonymous with $f(z) = O(g(z))$ $(z \to z_0)$, where O is Landau's asymptotic symbol. If the constant C depends on some parameters $\lambda, \mu, \ldots$, we may write $f(z) \ll_{\lambda,\mu,\ldots} g(z)$, or $f(z) = O_{\lambda,\mu,\ldots}(g(z))$.

Similarly, given two sequences $a_n \in \mathbb{C}$ and $b_n > 0$, we write

$$a_n \ll b_n$$

to mean $a_n = O(b_n)$, i.e., that there exists a constant $C > 0$ independent of n such that, for any n, $|a_n| \leq Cb_n$.

If $f(z)$ and $g(z)$ are both > 0, Vinogradov's notation $g(z) \gg f(z)$ means $f(z) \ll g(z)$, and similarly for sequences $a_n, b_n > 0$.

3.1 Order of a Function

We define the *order* of an entire function $F(z)$ as the infimum α of the exponents $A > 0$ such that

$$F(z) \ll \exp(|z|^A) \quad (z \to \infty), \tag{3.2}$$

and we denote

$$\alpha = \operatorname{ord} F(z).$$

C. Viola, *An Introduction to Special Functions*,
UNITEXT - La Matematica per il 3+2 102, DOI 10.1007/978-3-319-41345-7_3

If there exist no $A > 0$ satisfying (3.2), we define $\operatorname{ord} F(z) = +\infty$.

From the above definition it follows that

$$F(z) \ll_\varepsilon \exp(|z|^{A+\varepsilon}) \quad \text{for every } \varepsilon > 0 \tag{3.3}$$

if and only if $\operatorname{ord} F(z) \le A$.

Lemma 3.1 *If* $\operatorname{ord} F_1(z) = \alpha_1$ *and* $\operatorname{ord} F_2(z) = \alpha_2$, *then* $\operatorname{ord}(F_1(z) + F_2(z)) \le \max\{\alpha_1, \alpha_2\}$ *and* $\operatorname{ord}(F_1(z)F_2(z)) \le \max\{\alpha_1, \alpha_2\}$.

Proof Let $\alpha = \max\{\alpha_1, \alpha_2\}$. Then

$$F_1(z) \ll_\varepsilon \exp(|z|^{\alpha+\varepsilon}) \text{ and } F_2(z) \ll_\varepsilon \exp(|z|^{\alpha+\varepsilon}),$$

whence

$$|F_1(z) + F_2(z)| \le |F_1(z)| + |F_2(z)| \ll_\varepsilon \exp(|z|^{\alpha+\varepsilon})$$

and

$$|F_1(z)F_2(z)| = |F_1(z)|\,|F_2(z)| \ll_\varepsilon \exp(2|z|^{\alpha+\varepsilon}) \ll_\varepsilon \exp(|z|^{\alpha+2\varepsilon}). \qquad \square$$

Remark 3.1 Since $|z|^n \ll \exp(|z|^\varepsilon)$, from Lemma 3.1 it follows that, for any polynomial $P(z)$, $\operatorname{ord} P(z) = 0$. If $\deg P(z) = n$, then $\operatorname{ord} e^{P(z)} = n$. To see this, note that $|P(z)| \ll |z|^n$, whence $\left|e^{P(z)}\right| = e^{\operatorname{Re} P(z)} \le e^{|P(z)|} \ll \exp(|z|^{n+\varepsilon})$, so that $\operatorname{ord} e^{P(z)} \le n$. If $P(z) = a_0 z^n + a_1 z^{n-1} + \ldots + a_n$ ($a_0 \ne 0$), and if $z \to \infty$ along one of the n half-lines defined by $\arg z = -\frac{1}{n}\arg a_0$, whence $\arg(a_0 z^n) = \arg a_0 + n \arg z = 0$ so that $a_0 z^n > 0$, then $\operatorname{Re} P(z) - |z|^{n-\varepsilon} = \operatorname{Re}(a_0 z^n) + O(|z|^{n-\varepsilon}) = |a_0||z|^n + O(|z|^{n-\varepsilon}) \to +\infty$. This shows that

$$\limsup_{z\to\infty} \frac{\left|e^{P(z)}\right|}{\exp(|z|^{n-\varepsilon})} = \limsup_{z\to\infty} \exp(\operatorname{Re} P(z) - |z|^{n-\varepsilon}) = +\infty,$$

whence $\operatorname{ord} e^{P(z)} \ge n$.

We aim at relating the order of an entire function with the density of its zeros. To do this, we require the following

Lemma 3.2 *Let* $0 < r < R$, *and let* $f(z)$ *be regular in the disc* $|z - z_0| \le R$ *and not identically zero. Let* N *be the number of zeros of* $f(z)$ *in the disc* $|z - z_0| \le r$, *counted with multiplicity. Then*

$$|f(z_0)| \le \left(\frac{r}{R}\right)^N \max_{|z-z_0|=R} |f(z)|, \tag{3.4}$$

whence, if $f(z_0) \ne 0$,

$$N \le \frac{1}{\log(R/r)} \log \frac{\max_{|z-z_0|=R} |f(z)|}{|f(z_0)|}. \tag{3.5}$$

Proof The inequality (3.5) is an obvious consequence of (3.4) by taking logarithms. In order to prove (3.4), by a translation we may assume $z_0 = 0$. Also, by replacing $f(z)$ with $\widetilde{f}(z) := f(Rz)$, we may assume $R = 1$, since $\widetilde{f}(z)$ is regular in $|z| \le 1$ and has N zeros in $|z| \le r/R$. Let $z_1, \dots, z_N$ be the zeros of $f(z)$ in $|z| \le r$, where a zero of multiplicity m is repeated m times, and let

$$g(z) = f(z) \prod_{n=1}^{N} \frac{1 - \overline{z_n}\, z}{z - z_n}.$$

Since $f(z_n) = 0$, $g(z)$ is regular in $|z| \le 1$. Moreover, each of the Blaschke factors $(1 - \overline{z_n}\, z)/(z - z_n)$ maps the circumference $|z| = 1$ onto itself, because for any ζ with $|\zeta| \ne 1$ we get

$$\left| \frac{1 - \overline{\zeta}\, e^{i\vartheta}}{e^{i\vartheta} - \zeta} \right| = \left| e^{i\vartheta} \right| \left| \frac{e^{-i\vartheta} - \overline{\zeta}}{e^{i\vartheta} - \zeta} \right| = 1.$$

Thus, for $|t| \le 1$,

$$|g(t)| \le \max_{|z|=1} |g(z)| = \max_{|z|=1} |f(z)|,$$

whence

$$|f(t)| = |g(t)| \prod_{n=1}^{N} \left| \frac{t - z_n}{1 - \overline{z_n}\, t} \right| \le \left(\max_{|z|=1} |f(z)| \right) \prod_{n=1}^{N} \left| \frac{t - z_n}{1 - \overline{z_n}\, t} \right|.$$

In particular, for $t = 0$,

$$|f(0)| \le \left(\max_{|z|=1} |f(z)| \right) \prod_{n=1}^{N} |z_n| \le r^N \max_{|z|=1} |f(z)|.$$

This proves (3.4). □

Theorem 3.1 *Let $F(z)$ be an entire function not identically zero, and let $N(r)$ denote the number of zeros of $F(z)$ in the disc $|z| \le r$, counted with multiplicity. If*

$$\alpha = \operatorname{ord} F(z) < +\infty,$$

then, for any $\varepsilon > 0$,

$$N(r) \ll_\varepsilon r^{\alpha+\varepsilon} \quad (r \to +\infty). \tag{3.6}$$

Proof Assume $F(0) \ne 0$. By (3.5) with $z_0 = 0$, $R = 2r$, we obtain

$$N(r) \le \frac{1}{\log 2} \log \frac{\max_{|z|=2r} |F(z)|}{|F(0)|}.$$

Since

$$\max_{|z|=2r} |F(z)| \ll \exp\left((2r)^{\alpha+\varepsilon}\right),$$

we get

$$N(r) \ll (2r)^{\alpha+\varepsilon} \ll r^{\alpha+\varepsilon}.$$

If $z = 0$ is a zero of $F(z)$ of multiplicity m, we apply (3.6) to the function $z^{-m}F(z)$ which plainly satisfies $\operatorname{ord}\left(z^{-m}F(z)\right) = \alpha$. □

Let $z_1, z_2, \ldots \to \infty$ be a sequence of non-zero complex numbers. We define the *exponent of convergence* of the sequence (z_n) as the infimum β of the exponents $B > 0$ such that

$$\sum_{n=1}^{\infty} |z_n|^{-B} < +\infty. \tag{3.7}$$

If $\sum\limits_{n=1}^{\infty} |z_n|^{-B} = +\infty$ for every $B > 0$, the exponent of convergence of (z_n) is $+\infty$.

Theorem 3.2 *If an entire function $F(z)$ satisfying $F(0) \neq 0$ has order α, and if the sequence $z_1, z_2, \ldots$ of the zeros of $F(z)$ has exponent of convergence β, then $\beta \leq \alpha$.*

Proof From (3.6) with $r = |z_n|$ we get $n \ll |z_n|^{\alpha+\varepsilon}$, whence

$$|z_n|^{-(\alpha+2\varepsilon)} \ll n^{-(\alpha+2\varepsilon)/(\alpha+\varepsilon)} = n^{-(1+\varepsilon_1)},$$

for a suitable $\varepsilon_1 > 0$. Therefore

$$\sum_{n=1}^{\infty} |z_n|^{-(\alpha+2\varepsilon)} \ll \sum_{n=1}^{\infty} n^{-(1+\varepsilon_1)} < +\infty.$$

By (3.7) we get $\beta \leq \alpha$. □

Theorem 3.3 *If the zeros $z_1, z_2, \ldots$ of an entire function $F(z)$ such that $F(0) \neq 0$ have exponent of convergence $\beta < +\infty$, the Weierstrass factorization formula (2.13) can be written in the form*

$$F(z) = e^{G(z)} \prod_n \left(1 - \frac{z}{z_n}\right) \exp \sum_{k=1}^{p} \frac{1}{k}\left(\frac{z}{z_n}\right)^k, \tag{3.8}$$

where $p \geq 0$ is an integer independent of n.

Proof For any integer p such that $p + 1 > \beta$ we get

$$\sum_n |z_n|^{-(p+1)} < +\infty, \tag{3.9}$$

whence, for every $z \in \mathbb{C}$,

$$\sum_n \left|\frac{z}{z_n}\right|^{p+1} < +\infty,$$

i.e., the constant sequence $p_n = p$ satisfies (2.18). □

For a sequence $z_1, z_2, \ldots \to \infty$ of non-zero complex numbers with exponent of convergence $\beta < +\infty$, we use the standard notation

$$E\Big(\frac{z}{z_n}, p\Big) := \Big(1 - \frac{z}{z_n}\Big) \exp \sum_{k=1}^{p} \frac{1}{k}\Big(\frac{z}{z_n}\Big)^k, \tag{3.10}$$

and, if p is the least integer satisfying (3.9), we call

$$\prod_n E\Big(\frac{z}{z_n}, p\Big) = \prod_n \Big(1 - \frac{z}{z_n}\Big) \exp \sum_{k=1}^{p} \frac{1}{k}\Big(\frac{z}{z_n}\Big)^k \tag{3.11}$$

the Weierstrass *canonical product* for the sequence (z_n).

3.2 Hadamard's Theorem

In accordance with Theorems 3.2 and 3.3, for any entire function $F(z)$ satisfying $F(0) \neq 0$ and $\alpha = \operatorname{ord} F(z) < +\infty$ we shall use the factorization formula (3.8) with the canonical product (3.11). For such a function $F(z)$, the following theorem yields strong information on the entire function $G(z)$ appearing in (3.8).

Theorem 3.4 (Hadamard) *Let*

$$F(z) = e^{G(z)} \prod_n \Big(1 - \frac{z}{z_n}\Big) \exp \sum_{k=1}^{p} \frac{1}{k}\Big(\frac{z}{z_n}\Big)^k \tag{3.12}$$

be an entire function of order $\alpha < +\infty$, where the product is the canonical product for the zeros z_n of $F(z)$. Then the entire function $G(z)$ in (3.12) is a polynomial of degree not exceeding α.

Proof Let $\nu = \lfloor \alpha \rfloor$ (the integer part of α). By the Taylor series expansion of $G(z)$, the theorem is equivalent to proving that $G^{(\nu+1)}(z) = 0$ identically. Since p is the least integer satisfying (3.9) we have $p \leq \beta$, where β is the exponent of convergence of the sequence (z_n). By Theorem 3.2 we get $p \leq \alpha$, whence $p \leq \nu = \lfloor \alpha \rfloor$. Therefore

$$\frac{\mathrm{d}^{\nu+1}}{\mathrm{d}z^{\nu+1}} \sum_{k=1}^{p} \frac{1}{k}\Big(\frac{z}{z_n}\Big)^k = 0.$$

Thus, taking the logarithm of (3.12) and then differentiating $\nu + 1$ times we obtain

$$\frac{\mathrm{d}^\nu}{\mathrm{d}z^\nu}\frac{F'(z)}{F(z)} = G^{(\nu+1)}(z) - \nu! \sum_n \frac{1}{(z_n - z)^{\nu+1}}, \tag{3.13}$$

because term-by-term differentiation is allowed by (2.15) and by uniform convergence of the Mittag-Leffler series (2.3).

For any $R > 0$ we define

$$\varphi_R(z) := \frac{F(z)}{F(0)} \prod_{|z_n| \le R} \left(1 - \frac{z}{z_n}\right)^{-1}, \tag{3.14}$$

and

$$\psi_R(z) := \log \varphi_R(z), \tag{3.15}$$

where we take the principal logarithm at $z = 0$, i.e., $\psi_R(0) = \log 1 = 0$. Plainly $\varphi_R(z)$ is an entire function, and satisfies $\varphi_R(z) \neq 0$ for $|z| \le R$. Thus $\psi_R(z)$ is regular in the disc $|z| \le R$. Also, from (3.14) and (3.15) we get

$$\psi_R(z) = \log F(z) - \log F(0) - \sum_{|z_n| \le R} \log(z_n - z) + \sum_{|z_n| \le R} \log z_n.$$

Therefore

$$\psi_R^{(\nu+1)}(z) = \frac{\mathrm{d}^\nu}{\mathrm{d}z^\nu}\frac{F'(z)}{F(z)} + \nu! \sum_{|z_n| \le R} \frac{1}{(z_n - z)^{\nu+1}},$$

and (3.13) yields

$$\begin{aligned} G^{(\nu+1)}(z) &= \frac{\mathrm{d}^\nu}{\mathrm{d}z^\nu}\frac{F'(z)}{F(z)} + \nu! \sum_n \frac{1}{(z_n - z)^{\nu+1}} \\ &= \psi_R^{(\nu+1)}(z) + \nu! \sum_{|z_n| > R} \frac{1}{(z_n - z)^{\nu+1}}. \end{aligned} \tag{3.16}$$

Since $|1 - z/z_n| \ge |z|/|z_n| - 1 \ge 1$ for $|z| = 2R$ and $|z_n| \le R$, we get by (3.14)

$$|\varphi_R(z)| \le \frac{|F(z)|}{|F(0)|} \ll_\varepsilon \exp\left((2R)^{\alpha+\varepsilon}\right) \tag{3.17}$$

for $|z| = 2R$, hence also for $|z| \le 2R$ by the maximum principle. From (3.15) and (3.17) we obtain

$$\operatorname{Re} \psi_R(z) = \log |\varphi_R(z)| \ll_\varepsilon R^{\alpha+\varepsilon} \quad \text{for } |z| \le R. \tag{3.18}$$

Moreover, again by the maximum principle, from $\psi_R(0) = 0$ we get

$$\max_{|z|=R} \operatorname{Re} \psi_R(z) \geq 0.$$

Thus we can apply Borel–Carathéodory's inequality (1.2) to the function $\psi_R(z)$ with $r = R/2$ and $z_0 = 0$. By (3.18) we obtain

$$\max_{|z|=R/2} \left|\psi_R^{(\nu+1)}(z)\right| \ll_\nu \frac{R}{(R/2)^{\nu+2}} \max_{|z|=R} \operatorname{Re} \psi_R(z) \ll_{\varepsilon,\nu} R^{\alpha-\nu-1+\varepsilon}.$$

Hence for $|z| \leq R/2$ we get, by (3.16),

$$\begin{aligned} G^{(\nu+1)}(z) &\ll_{\varepsilon,\nu} R^{\alpha-\nu-1+\varepsilon} + \sum_{|z_n|>R} \frac{1}{|z_n - z|^{\nu+1}} \\ &\ll_{\varepsilon,\nu} R^{\alpha-\nu-1+\varepsilon} + \sum_{|z_n|>R} |z_n|^{-(\nu+1)}, \end{aligned} \tag{3.19}$$

because $|z_n - z| \geq |z_n| - |z| \geq |z_n| - R/2 > \frac{1}{2}|z_n|$. Since $\nu = \lfloor \alpha \rfloor$ we have $\alpha < \nu + 1$, whence

$$\lim_{R\to\infty} R^{\alpha-\nu-1+\varepsilon} = 0$$

for sufficiently small $\varepsilon > 0$. Moreover, by Theorem 3.2, $\beta \leq \alpha < \nu + 1$, whence $\sum_n |z_n|^{-(\nu+1)} < +\infty$. Thus

$$\lim_{R\to\infty} \sum_{|z_n|>R} |z_n|^{-(\nu+1)} = 0.$$

For any fixed $z \in \mathbb{C}$, (3.19) holds for all $R \geq 2|z|$. Since $G^{(\nu+1)}(z)$ is independent of R, making $R \to \infty$ in (3.19) we obtain $G^{(\nu+1)}(z) = 0$. □

For any polynomial $P(z)$ of degree n, $e^{P(z)}$ is a non-vanishing entire function, of order n by Remark 3.1 above. The converse of this statement is also true, in the following form.

Corollary 3.1 *Let $F(z)$ be an entire function satisfying $F(z) \neq 0$ for all $z \in \mathbb{C}$, and let $\alpha = \operatorname{ord} F(z) < +\infty$. Then $\alpha \in \mathbb{N}$ and*

$$F(z) = e^{G(z)},$$

where $G(z)$ is a polynomial of degree α.

Proof We use a simplified version of the proof of Theorem 3.4, where the product $\prod_n$ on the right-hand side of (3.12) is empty. Let $\nu = \lfloor \alpha \rfloor$, let

$$\psi(z) = \log \frac{F(z)}{F(0)},$$

where we take the principal logarithm at $z = 0$, i.e., $\psi(0) = \log 1 = 0$, and let

$$G(z) = \log F(z) = \psi(z) + \log F(0).$$

Then

$$G^{(\nu+1)}(z) = \psi^{(\nu+1)}(z) = \frac{\mathrm{d}^\nu}{\mathrm{d}z^\nu} \frac{F'(z)}{F(z)}. \tag{3.20}$$

Since $F(z) \neq 0$ for all $z \in \mathbb{C}$, $G(z)$ and $\psi(z)$ are entire functions. For any $R > 0$, by the maximum principle we have

$$\max_{|z|=R} \operatorname{Re} \psi(z) \geq \operatorname{Re} \psi(0) = 0.$$

Thus, by Borel–Carathéodory's inequality (1.2) with $r = R/2$ and $z_0 = 0$,

$$\max_{|z|=R/2} \left|\psi^{(\nu+1)}(z)\right| \ll \frac{R}{(R/2)^{\nu+2}} \max_{|z|=R} \operatorname{Re} \psi(z). \tag{3.21}$$

Since $\alpha = \operatorname{ord} F(z)$, for $|z| \leq R$ and for any $\varepsilon > 0$ we get

$$\left|\frac{F(z)}{F(0)}\right| \ll_\varepsilon \exp\left(R^{\alpha+\varepsilon}\right),$$

whence

$$\operatorname{Re} \psi(z) = \log \left|\frac{F(z)}{F(0)}\right| \ll_\varepsilon R^{\alpha+\varepsilon}.$$

Therefore, by (3.20) and (3.21),

$$\max_{|z|\leq R/2} \left|G^{(\nu+1)}(z)\right| = \max_{|z|\leq R/2} \left|\psi^{(\nu+1)}(z)\right| \ll_\varepsilon R^{\alpha-\nu-1+\varepsilon}.$$

Hence, for any fixed $z \in \mathbb{C}$,

$$\left|G^{(\nu+1)}(z)\right| \ll_\varepsilon R^{\alpha-\nu-1+\varepsilon} \quad \text{for all } R \geq 2|z|.$$

Since $\nu = \lfloor\alpha\rfloor$ we have $\alpha < \nu + 1$, whence $\alpha - \nu - 1 + \varepsilon < 0$ for sufficiently small $\varepsilon > 0$. For $R \to \infty$ we obtain $G^{(\nu+1)}(z) = 0$ identically. The Taylor expansion of $G(z)$ implies that $G(z)$ is a polynomial of degree $\mu \leq \nu = \lfloor\alpha\rfloor$. By Remark 3.1, $F(z) = e^{G(z)}$ has order μ, whence $\alpha = \mu \leq \lfloor\alpha\rfloor$. Thus $\alpha = \mu = \lfloor\alpha\rfloor \in \mathbb{N}$, and $\deg G(z) = \mu = \alpha$. □

By (2.15) and Lemma 2.2 the Weierstrass canonical product (3.11) is an entire function, the order of which is given by the following

Theorem 3.5 *Let $z_1, z_2, \ldots \to \infty$ be any sequence of non-zero complex numbers with exponent of convergence $\beta < +\infty$. The canonical product*

$$\prod_n E\Big(\frac{z}{z_n}, p\Big) = \prod_n \Big(1 - \frac{z}{z_n}\Big) \exp \sum_{k=1}^{p} \frac{1}{k}\Big(\frac{z}{z_n}\Big)^k$$

is an entire function of order β.

Proof Let α be the order of the canonical product (3.11). By Theorem 3.2 we know that $\beta \le \alpha$. Thus we must prove that $\alpha \le \beta$, i.e., by (3.3),

$$\prod_n E\Big(\frac{z}{z_n}, p\Big) \ll_\varepsilon \exp(|z|^{\beta+\varepsilon})$$

for every $\varepsilon > 0$. For this purpose, it suffices to prove that

$$\log\Big|\prod_n E\Big(\frac{z}{z_n}, p\Big)\Big| = \sum_n \log\Big|E\Big(\frac{z}{z_n}, p\Big)\Big| \ll_\varepsilon |z|^{\beta+\varepsilon} \quad (z \to \infty), \qquad (3.22)$$

because from (3.22) with $\varepsilon/2$ in place of ε one gets

$$\log\Big|\prod_n E\Big(\frac{z}{z_n}, p\Big)\Big| \ll_\varepsilon |z|^{\beta+\varepsilon/2} = o(|z|^{\beta+\varepsilon}) \le \vartheta|z|^{\beta+\varepsilon}$$

with $0 < \vartheta < 1$, whence

$$\Big|\prod_n E\Big(\frac{z}{z_n}, p\Big)\Big| \le \exp(\vartheta|z|^{\beta+\varepsilon}) = o\big(\exp(|z|^{\beta+\varepsilon})\big).$$

If $|z/z_n| \le \frac{1}{2}$ we have

$$\begin{aligned}\log\Big|E\Big(\frac{z}{z_n}, p\Big)\Big| &\le \Big|\log E\Big(\frac{z}{z_n}, p\Big)\Big| = \Big|\log\Big(1 - \frac{z}{z_n}\Big) + \sum_{k=1}^{p} \frac{1}{k}\Big(\frac{z}{z_n}\Big)^k\Big| \\ &= \Big|-\sum_{k=p+1}^{\infty} \frac{1}{k}\Big(\frac{z}{z_n}\Big)^k\Big| \le \sum_{k=p+1}^{\infty} \Big|\frac{z}{z_n}\Big|^k \\ &\le \Big|\frac{z}{z_n}\Big|^{p+1}\Big(1 + \frac{1}{2} + \frac{1}{2^2} + \ldots\Big) = 2\Big|\frac{z}{z_n}\Big|^{p+1},\end{aligned}$$

whence, by (3.9),

$$\sum_{|z_n| \ge 2|z|} \log\Big|E\Big(\frac{z}{z_n}, p\Big)\Big| \le 2\,|z|^{p+1} \sum_{|z_n| \ge 2|z|} |z_n|^{-(p+1)} \ll |z|^{p+1}. \qquad (3.23)$$

Since p is the least integer satisfying (3.9), we have $p \le \beta \le p+1$. If $\beta = p+1$, (3.23) yields

$$\sum_{|z_n|\ge 2|z|} \log\left|E\left(\frac{z}{z_n},\, p\right)\right| \ll |z|^{\beta}.$$

If $\beta < p+1$, for any ε satisfying $0 < \varepsilon < p+1-\beta$ we get by (3.23)

$$\begin{aligned}\sum_{|z_n|\ge 2|z|} \log\left|E\left(\frac{z}{z_n},\, p\right)\right| &\le 2\,|z|^{\beta+\varepsilon} \sum_{|z_n|\ge 2|z|} |z_n|^{-(p+1)}\,|z|^{p+1-\beta-\varepsilon} \\ &\le \frac{2\,|z|^{\beta+\varepsilon}}{2^{p+1-\beta-\varepsilon}} \sum_{|z_n|\ge 2|z|} |z_n|^{-(\beta+\varepsilon)} \ll_{\varepsilon} |z|^{\beta+\varepsilon}.\end{aligned}$$

In either case

$$\sum_{|z_n|\ge 2|z|} \log\left|E\left(\frac{z}{z_n},\, p\right)\right| \ll_{\varepsilon} |z|^{\beta+\varepsilon}. \tag{3.24}$$

If $|z/z_n| > \frac{1}{2}$ we have

$$\begin{aligned}\log\left|E\left(\frac{z}{z_n},\, p\right)\right| &= \log\left|1-\frac{z}{z_n}\right| + \operatorname{Re}\sum_{k=1}^{p}\frac{1}{k}\left(\frac{z}{z_n}\right)^k \\ &\le \log\left(1+\left|\frac{z}{z_n}\right|\right) + \sum_{k=1}^{p}\left|\frac{z}{z_n}\right|^k \ll \begin{cases}\left|\dfrac{z}{z_n}\right|^{\varepsilon} & \text{if } p=0 \\ \left|\dfrac{z}{z_n}\right|^{p} & \text{if } p\ge 1\end{cases} \\ &\ll_{\varepsilon} \left|\frac{z}{z_n}\right|^{p+\varepsilon}.\end{aligned}$$

Therefore

$$\begin{aligned}\sum_{|z_n|<2|z|} \log\left|E\left(\frac{z}{z_n},\, p\right)\right| &\ll_{\varepsilon} |z|^{p+\varepsilon} \sum_{|z_n|<2|z|} |z_n|^{-(p+\varepsilon)} \\ &= |z|^{\beta+\varepsilon} \sum_{|z_n|<2|z|} |z_n|^{-(p+\varepsilon)}\,|z|^{-(\beta-p)} \\ &\le 2^{\beta-p}\,|z|^{\beta+\varepsilon} \sum_{|z_n|<2|z|} |z_n|^{-(\beta+\varepsilon)} \ll_{\varepsilon} |z|^{\beta+\varepsilon}.\end{aligned} \tag{3.25}$$

From (3.24) and (3.25) we obtain

$$\sum_{n} \log\left|E\left(\frac{z}{z_n},\, p\right)\right| = \sum_{|z_n|\ge 2|z|} \log\left|E\left(\frac{z}{z_n},\, p\right)\right| + \sum_{|z_n|<2|z|} \log\left|E\left(\frac{z}{z_n},\, p\right)\right| \ll_{\varepsilon} |z|^{\beta+\varepsilon},$$

and (3.22) is proved. □

We remark that Theorem 3.5 implies the existence of entire functions of any real order $\alpha \geq 0$. The polynomials are entire functions of order 0 by Remark 3.1. For any $\alpha > 0$, the sequence $n^{1/\alpha}$ $(n = 1, 2, \ldots)$ has exponent of convergence α, whence the canonical product (3.11) for the sequence $z_n = n^{1/\alpha}$ is an entire function of order α by Theorem 3.5.

Corollary 3.2 *Let*

$$F(z) = e^{G(z)} \prod_n \left(1 - \frac{z}{z_n}\right) \exp \sum_{k=1}^{p} \frac{1}{k} \left(\frac{z}{z_n}\right)^k$$

be an entire function of order α*, where the product is the canonical product for the zeros* z_n *of* $F(z)$*. Let* β *be the exponent of convergence of the sequence* (z_n) *and let* $G(z)$ *be a polynomial of degree* q*. Then*

$$\alpha = \operatorname{ord} F(z) = \max\{\beta, q\}.$$

Proof We have $\beta \leq \alpha$ by Theorem 3.2 and $q \leq \alpha$ by Theorem 3.4, whence

$$\max\{\beta, q\} \leq \alpha.$$

Also, $\operatorname{ord} e^{G(z)} = q$ by Remark 3.1 and $\operatorname{ord} \prod_n E(z/z_n, p) = \beta$ by Theorem 3.5. From Lemma 3.1 we get

$$\alpha = \operatorname{ord}\left(e^{G(z)} \prod_n E\left(\frac{z}{z_n}, p\right)\right) \leq \max\{\beta, q\}. \qquad \square$$

The *genus* g of the function

$$F(z) = e^{G(z)} \prod_n E\left(\frac{z}{z_n}, p\right)$$

is defined to be

$$g = \max\{p, q\},$$

where p is the least integer satisfying (3.9) and q is the degree of the polynomial $G(z)$. Since $p \leq \beta \leq p + 1$, for the integer g we get, by Corollary 3.2,

$$g \leq \alpha \leq g + 1.$$

Corollary 3.3 *Under the assumptions of Corollary 3.2, the polynomial* $G(z)$ *is given by*

$$G(z) = \log F(0) + \sum_{k=1}^{p} \frac{z^k}{k!} \left[\frac{\mathrm{d}^{k-1}}{\mathrm{d}z^{k-1}} \frac{F'(z)}{F(z)}\right]_{z=0} \quad \text{if } q \leq p, \tag{3.26}$$

or by

$$G(z) = \log F(0) + \sum_{k=1}^{q} \frac{z^k}{k!}\left[\frac{\mathrm{d}^{k-1}}{\mathrm{d}z^{k-1}}\frac{F'(z)}{F(z)}\right]_{z=0} + \sum_{k=p+1}^{q} \frac{z^k}{k} \sum_n z_n^{-k} \quad \text{if } q > p. \tag{3.27}$$

Moreover, for any integer $k > g = \max\{p, q\}$ *we get*

$$\sum_n z_n^{-k} = -\frac{1}{(k-1)!}\left[\frac{\mathrm{d}^{k-1}}{\mathrm{d}z^{k-1}}\frac{F'(z)}{F(z)}\right]_{z=0}. \tag{3.28}$$

Proof Since $G(z)$ is a polynomial of degree q, (2.17) yields (3.26) and (3.27). Also, for any $k > q$ the coefficient of z^k in (2.17) vanishes. Thus for $k > \max\{p, q\}$ we get

$$\frac{1}{(k-1)!}\left[\frac{\mathrm{d}^{k-1}}{\mathrm{d}z^{k-1}}\frac{F'(z)}{F(z)}\right]_{z=0} + \sum_n z_n^{-k} = 0,$$

whence (3.28). □

Chapter 4
Bernoulli Numbers and Polynomials

4.1 Euler's Factorization of sin z

By a standard application of the Weierstrass–Hadamard factorization formula (3.12) we prove the following

Theorem 4.1 (Euler) *The entire function* $\sin(\pi z)/\pi z$ *satisfies*

$$\frac{\sin(\pi z)}{\pi z} = \prod_{n=1}^{\infty}\left(1 - \frac{z^2}{n^2}\right) \tag{4.1}$$

for all $z \in \mathbb{C}$.

Proof By Remark 3.1, the exponential functions $e^{\pm\pi i z}$ have order 1. Thus, by Lemma 3.1, the entire function

$$F(z) := \frac{\sin(\pi z)}{\pi z} = \frac{1}{2\pi i z}(e^{\pi i z} - e^{-\pi i z}) = 1 - \frac{\pi^2 z^2}{3!} + \dots$$

has order $\alpha \leq 1$ and satisfies

$$F(0) = 1, \quad F'(0) = 0. \tag{4.2}$$

Since $e^{\pi i z} = e^{-\pi i z}$ if and only if $e^{2\pi i z} = 1$, i.e., $2\pi i z = \log 1 = 2k\pi i$ with $k \in \mathbb{Z}$, the zeros of $F(z)$ are $z = \pm 1, \pm 2, \pm 3, \dots$ with exponent of convergence $\beta = 1$. By Theorem 3.2 we have $\beta = 1 \leq \alpha$, whence $\alpha = \beta = 1$. Also, the least integer p satisfying (3.9) for $|z_n| = n$ is plainly $p = 1$. Hence, in the present case, (3.12) becomes

$$F(z) = \frac{\sin(\pi z)}{\pi z} = e^{G(z)}\prod_{n=1}^{\infty}\left(1 - \frac{z}{n}\right)e^{z/n}\left(1 + \frac{z}{n}\right)e^{-z/n} = e^{G(z)}\prod_{n=1}^{\infty}\left(1 - \frac{z^2}{n^2}\right),$$

C. Viola, *An Introduction to Special Functions*,
UNITEXT - La Matematica per il 3+2 102, DOI 10.1007/978-3-319-41345-7_4

where $G(z)$ is a polynomial of degree $q \leq \alpha = p = 1$ by Theorem (3.4). Thus, by (3.26) and (4.2),

$$G(z) = \log F(0) + \frac{F'(0)}{F(0)} z = 0,$$

and (4.1) follows. □

Since the zeros of the entire function

$$\cos(\pi z) = \frac{1}{2}(e^{\pi i z} + e^{-\pi i z}) = 1 - \frac{\pi^2 z^2}{2!} + \dots$$

are $z = \pm(n - 1/2)$ $(n = 1, 2, 3, \dots)$, the same method used in the proof of Theorem 4.1 yields the factorization

$$\cos(\pi z) = \prod_{n=1}^{\infty}\left(1 - \frac{4z^2}{(2n-1)^2}\right).$$

The Riemann zeta-function is defined by

$$\zeta(s) := \sum_{n=1}^{\infty} n^{-s} \qquad (s \in \mathbb{C},\ \mathrm{Re}\, s > 1). \tag{4.3}$$

As Riemann proved in his celebrated paper, published in 1859, on the average distribution of prime numbers, $\zeta(s)$ can be analytically continued in the whole $\mathbb{C}$ to a meromorphic function of s, regular in $\mathbb{C} \setminus \{1\}$ and having a simple pole with residue 1 at $s = 1$. Here our aim is to prove Euler's formulae (4.16) for the values

$$\zeta(2k) = \sum_{n=1}^{\infty} n^{-2k} \qquad (k = 1, 2, 3, \dots) \tag{4.4}$$

of the zeta-function at even positive integers $2k$. We apply (3.28) to the function (4.1). The logarithmic derivative of (4.1) is

$$\frac{\mathrm{d}}{\mathrm{d}z}\big(\log \sin(\pi z) - \log z - \log \pi\big) = \pi \cot(\pi z) - \frac{1}{z},$$

and the genus of (4.1) is $g = \max\{p, q\} = 1$. Thus (3.28) yields

$$\sum_{n=1}^{\infty} n^{-h} + \sum_{n=1}^{\infty}(-n)^{-h} = -\frac{1}{(h-1)!}\left[\frac{\mathrm{d}^{h-1}}{\mathrm{d}z^{h-1}}\left(\pi \cot(\pi z) - \frac{1}{z}\right)\right]_{z=0}$$

$$(h = 2, 3, 4, \dots). \tag{4.5}$$

For odd $h = 2k + 1$ the two sides of (4.5) vanish; hence (4.5) gives no information on $\zeta(2k+1)$. For even $h = 2k$, (4.5) becomes

$$\zeta(2k) = \sum_{n=1}^{\infty} n^{-2k} = \frac{1}{(2k-1)!}\left[\frac{\mathrm{d}^{2k-1}}{\mathrm{d}z^{2k-1}}\left(\frac{1}{2z} - \frac{\pi}{2}\cot(\pi z)\right)\right]_{z=0} \quad (k = 1, 2, 3, \dots). \quad (4.6)$$

This formula is equivalent to the Taylor series expansion

$$\frac{1}{2z} - \frac{\pi}{2}\cot(\pi z) = \sum_{k=1}^{\infty} \zeta(2k)\, z^{2k-1} \qquad (|z| < 1). \quad (4.7)$$

4.2 Bernoulli Numbers

We can express (4.6) in a different form by means of the Bernoulli numbers, which allow to write the Laurent series expansion of $\cot z$ at $z = 0$.

Definition 4.1 The Bernoulli numbers are defined by

$$B_n = \left[\frac{\mathrm{d}^n}{\mathrm{d}z^n}\frac{z}{e^z - 1}\right]_{z=0} \qquad (n = 0, 1, 2, \dots),$$

i.e.,

$$\frac{z}{e^z - 1} = \sum_{n=0}^{\infty} B_n \frac{z^n}{n!} \qquad (|z| < 2\pi). \quad (4.8)$$

Since we have identically

$$\frac{z}{e^z - 1} + \frac{z}{2} = \frac{-z}{e^{-z} - 1} - \frac{z}{2},$$

the function $z/(e^z - 1) + z/2$ is even and takes the value 1 at $z = 0$. Hence (4.8) yields

$$B_0 = 1, \quad B_1 = -\frac{1}{2}, \quad B_{2k+1} = 0 \ \ (k = 1, 2, 3, \dots). \quad (4.9)$$

From (4.8) we get, for $|z| < 2\pi$,

$$1 = \left(\sum_{n=0}^{\infty} B_n \frac{z^n}{n!}\right) \frac{e^z - 1}{z} = \left(\sum_{n=0}^{\infty} B_n \frac{z^n}{n!}\right)\left(\sum_{n=1}^{\infty} \frac{z^{n-1}}{n!}\right)$$
$$= \sum_{n=1}^{\infty}\left(\sum_{k=0}^{n-1} \frac{B_k}{k!\,(n-k)!}\right) z^{n-1}.$$

Thus, for $n \geq 2$,

$$\sum_{k=0}^{n-1} \frac{B_k}{k!\,(n-k)!} = 0,$$

whence, multiplying by $n!$,

$$\sum_{k=0}^{n-1} \binom{n}{k} B_k = 0 \qquad (n = 2, 3, 4, \ldots). \tag{4.10}$$

Adding B_n, (4.10) yields

$$\sum_{k=0}^{n} \binom{n}{k} B_k = B_n \qquad (n = 0, 2, 3, 4, \ldots), \tag{4.11}$$

which can symbolically be written as

$$B_n = (1 + B)^n \qquad (n \neq 1),$$

where in the binomial expansion of $(1 + B)^n$ the symbolic powers B^k are replaced by B_k. Note that, for $n = 1$, instead of (4.11) we have, by (4.9),

$$\sum_{k=0}^{1} \binom{1}{k} B_k = \frac{1}{2} = -B_1. \tag{4.12}$$

From (4.10) we get the recurrence formula

$$B_{n-1} = -\frac{1}{n}\left(1 + \binom{n}{1} B_1 + \ldots + \binom{n}{n-2} B_{n-2}\right) \qquad (n = 2, 3, 4, \ldots), \tag{4.13}$$

which shows that $B_n \in \mathbb{Q}$ for all $n \geq 0$, and yields, for the first few B_{2k},

$$B_2 = \frac{1}{6}, \quad B_4 = -\frac{1}{30}, \quad B_6 = \frac{1}{42}, \quad B_8 = -\frac{1}{30},$$
$$B_{10} = \frac{5}{66}, \quad B_{12} = -\frac{691}{2730}, \quad B_{14} = \frac{7}{6}, \quad B_{16} = -\frac{3617}{510}, \ldots. \tag{4.14}$$

By (4.8) and (4.9), in the disc $|z| < \pi$ we get the Laurent expansion

$$\cot z = i\,\frac{e^{iz}+e^{-iz}}{e^{iz}-e^{-iz}} = i\,\frac{e^{2iz}+1}{e^{2iz}-1} = i + \frac{1}{z}\,\frac{2iz}{e^{2iz}-1} = i + \frac{1}{z}\sum_{n=0}^{\infty} B_n\,\frac{(2iz)^n}{n!}$$
$$= i + \frac{1}{z}\left(1 - iz + \sum_{k=1}^{\infty}(-1)^k B_{2k}\,\frac{(2z)^{2k}}{(2k)!}\right) = \frac{1}{z} + \sum_{k=1}^{\infty}(-1)^k 2^{2k} B_{2k}\,\frac{z^{2k-1}}{(2k)!},$$

whence

$$\frac{1}{2z} - \frac{\pi}{2}\cot(\pi z) = \sum_{k=1}^{\infty}(-1)^{k-1}(2\pi)^{2k} B_{2k}\,\frac{z^{2k-1}}{2\,(2k)!} \qquad (|z| < 1). \tag{4.15}$$

Comparing (4.7) with (4.15) we obtain Euler's formulae

$$\zeta(2k) = (-1)^{k-1}\,\frac{(2\pi)^{2k} B_{2k}}{2\,(2k)!} \qquad (k = 1, 2, 3, \ldots). \tag{4.16}$$

By (4.14) and (4.16),

$$\zeta(2) = \frac{\pi^2}{6}, \quad \zeta(4) = \frac{\pi^4}{90}, \quad \zeta(6) = \frac{\pi^6}{945}, \quad \zeta(8) = \frac{\pi^8}{9450}, \quad \ldots.$$

Euler's formulae (4.16) show that $(-1)^{k-1} B_{2k} > 0$ for $k \geq 1$, i.e., B_2, B_4, B_6, … have alternate signs. Moreover, writing (4.16) in the form

$$(-1)^{k-1} B_{2k} = \frac{2\,(2k)!}{(2\pi)^{2k}}\,\zeta(2k), \tag{4.17}$$

we get

$$2\,\frac{(2k)!}{(2\pi)^{2k}} < (-1)^{k-1} B_{2k} \leq \frac{\pi^2}{3}\,\frac{(2k)!}{(2\pi)^{2k}} \qquad (k = 1, 2, 3, \ldots), \tag{4.18}$$

because

$$1 < \zeta(2k) \leq \zeta(2) = \frac{\pi^2}{6}.$$

Since

$$\frac{\pi^2}{3}\,\frac{(2k)!}{(2\pi)^{2k}} \le 2\,\frac{(2k+2)!}{(2\pi)^{2k+2}}$$

if and only if $3(k+1)(2k+1) \ge \pi^4$, i.e., if and only if $k \ge 4$, by (4.18) we obtain $-B_8 < B_{10} < -B_{12} < \dots$. Thus from the values (4.14) we see that $|B_{2k}|$ decreases for $k \le 3$ and increases for $k \ge 3$. An asymptotic formula for the growth of $|B_{2k}|$ will be given in (5.13).

Since $B_{2k} \in \mathbb{Q}$, (4.16) shows that, for each $k = 1, 2, 3, \dots$, $\zeta(2k)$ is a rational multiple of π^{2k} and hence is transcendental by Lindemann's theorem (1882) on the transcendence of π.

For each $k = 1, 2, 3, \dots$, $\zeta(2k+1)$ is conjectured to be transcendental, but very little has been proved so far about the arithmetical nature of $\zeta(2k+1)$. Apéry (1979) proved that $\zeta(3) \notin \mathbb{Q}$, Rivoal (2000) proved that there exist infinitely many positive integers k such that $\zeta(2k+1) \notin \mathbb{Q}$ (although Rivoal's theorem is ineffective, i.e., for no explicitly given integer $k \ge 2$ the irrationality of $\zeta(2k+1)$ has been proved), and Zudilin (2004) proved that at least one of the four numbers $\zeta(5)$, $\zeta(7)$, $\zeta(9)$ and $\zeta(11)$ is irrational.

4.3 Bernoulli Polynomials

Definition 4.2 The Bernoulli polynomials are defined by

$$B_n(x) = \left[\frac{\partial^n}{\partial z^n}\,\frac{z\,e^{xz}}{e^z - 1}\right]_{z=0} \qquad (n = 0, 1, 2, \dots),$$

i.e.,

$$\frac{z\,e^{xz}}{e^z - 1} = \sum_{n=0}^{\infty} B_n(x)\,\frac{z^n}{n!} \qquad (|z| < 2\pi). \tag{4.19}$$

From the generating function (4.8) of the Bernoulli numbers we get, for $|z| < 2\pi$,

$$\frac{z\,e^{xz}}{e^z - 1} = \left(\sum_{n=0}^{\infty} B_n\,\frac{z^n}{n!}\right)\left(\sum_{n=0}^{\infty} x^n\,\frac{z^n}{n!}\right) = \sum_{n=0}^{\infty}\left(\sum_{k=0}^{n}\binom{n}{k} B_k\,x^{n-k}\right)\frac{z^n}{n!},$$

whence

$$B_n(x) = \sum_{k=0}^{n}\binom{n}{k} B_k\,x^{n-k} \qquad (n = 0, 1, 2, \dots), \tag{4.20}$$

which can symbolically be written as

$$B_n(x) = (x+B)^n.$$

Thus, by (4.20),

$$\begin{array}{ll} B_0(x) = 1, & B_1(x) = x - \frac{1}{2}, \\ B_2(x) = x^2 - x + \frac{1}{6}, & B_3(x) = x^3 - \frac{3}{2}x^2 + \frac{1}{2}x, \\ B_4(x) = x^4 - 2x^3 + x^2 - \frac{1}{30}, & B_5(x) = x^5 - \frac{5}{2}x^4 + \frac{5}{3}x^3 - \frac{1}{6}x, \\ \dots\dots\dots\dots & \dots\dots\dots\dots \end{array}$$

From (4.20) we get

$$B_n(0) = B_n \tag{4.21}$$

and

$$B_n(1) = \sum_{k=0}^{n} \binom{n}{k} B_k,$$

whence, by (4.9), (4.11) and (4.12),

$$B_n(1) = \begin{cases} B_n & \text{if } n \neq 1 \\ \frac{1}{2} & \text{if } n = 1 \end{cases} = (-1)^n B_n. \tag{4.22}$$

The Bernoulli polynomials satisfy several symmetry properties and functional equations, which are easily obtained from (4.19) or (4.20). By (4.19) we have, for any $x, y \in \mathbb{C}$ and for $|z| < 2\pi$,

$$\begin{aligned} \sum_{n=0}^{\infty} B_n(x+y)\frac{z^n}{n!} = \frac{z\,e^{xz}\,e^{yz}}{e^z - 1} &= \left(\sum_{n=0}^{\infty} B_n(x)\frac{z^n}{n!}\right)\left(\sum_{n=0}^{\infty} y^n \frac{z^n}{n!}\right) \\ &= \sum_{n=0}^{\infty}\left(\sum_{k=0}^{n}\binom{n}{k} B_k(x)\,y^{n-k}\right)\frac{z^n}{n!}, \end{aligned}$$

whence the identity

$$B_n(x+y) = \sum_{k=0}^{n}\binom{n}{k} B_k(x)\,y^{n-k}. \tag{4.23}$$

In particular, by (4.22),

$$B_n(1+x) = \sum_{k=0}^{n} \binom{n}{k} B_k(1)\, x^{n-k} = \sum_{\substack{k=0\\k\neq 1}}^{n} \binom{n}{k} B_k\, x^{n-k} + \frac{n}{2}\, x^{n-1}$$
$$= \sum_{k=0}^{n} \binom{n}{k} B_k\, x^{n-k} + n\, x^{n-1}.$$

Therefore, by (4.20),

$$B_n(x+1) - B_n(x) = n\, x^{n-1}. \tag{4.24}$$

Differentiating (4.24) we get, for $n \geq 1$,

$$B_n'(x+1) - B_n'(x) = n(n-1)\, x^{n-2} = n\big(B_{n-1}(x+1) - B_{n-1}(x)\big),$$

whence

$$B_n'(x+1) - n B_{n-1}(x+1) = B_n'(x) - n B_{n-1}(x).$$

Thus the polynomial $B_n'(x) - n B_{n-1}(x)$ has period 1, and therefore is constant. From (4.20) and (4.21) we obtain

$$B_n'(0) = \binom{n}{n-1} B_{n-1} = n B_{n-1}(0).$$

Hence $B_n'(x) - n B_{n-1}(x)$ is identically zero, and we get the differentiation formula

$$B_n'(x) = n B_{n-1}(x) \qquad (n = 1, 2, 3, \ldots). \tag{4.25}$$

Another interesting consequence of (4.24) is, for any $n, M, N \in \mathbb{N}$ with $M \leq N$,

$$B_{n+1}(N+1) - B_{n+1}(M) = \sum_{k=M}^{N} \big(B_{n+1}(k+1) - B_{n+1}(k)\big) = \sum_{k=M}^{N} (n+1)\, k^n,$$

whence the formula for the sum of consecutive nth powers:

$$\sum_{k=M}^{N} k^n = \frac{B_{n+1}(N+1) - B_{n+1}(M)}{n+1}. \tag{4.26}$$

For $M = 1$, (4.26) yields

$$
\begin{aligned}
1+2+\ldots+N &= \frac{1}{2}\left(B_2(N+1)-B_2\right) = \frac{1}{2}(N^2+N) = \frac{N(N+1)}{2},\\
1^2+2^2+\ldots+N^2 &= \frac{1}{3}\,B_3(N+1) = \frac{1}{3}\left(N^3+\frac{3}{2}N^2+\frac{1}{2}N\right)\\
&= \frac{N(N+1)(2N+1)}{6},\\
1^3+2^3+\ldots+N^3 &= \frac{1}{4}\left(B_4(N+1)-B_4\right) = \frac{1}{4}\left(N^4+2N^3+N^2\right)\\
&= \frac{N^2(N+1)^2}{4},\\
1^4+2^4+\ldots+N^4 &= \frac{1}{5}\,B_5(N+1) = \frac{1}{5}\left(N^5+\frac{5}{2}N^4+\frac{5}{3}N^3-\frac{1}{6}N\right)\\
&= \frac{N(N+1)(2N+1)(3N^2+3N-1)}{30},
\end{aligned}
$$

etc. Remarkably, from the above formulae we see that

$$1^3+2^3+\ldots+N^3 = (1+2+\ldots+N)^2.$$

From (4.22) and (4.23) we get

$$
\begin{aligned}
B_n(1-x) &= \sum_{k=0}^{n}\binom{n}{k}B_k(1)(-x)^{n-k} = \sum_{k=0}^{n}\binom{n}{k}(-1)^k B_k\cdot(-1)^{n-k}x^{n-k}\\
&= (-1)^n\sum_{k=0}^{n}\binom{n}{k}B_k\,x^{n-k},
\end{aligned}
$$

and (4.20) yields the symmetry property of $B_n(x)$ around $x = 1/2$:

$$B_n(1-x) = (-1)^n B_n(x). \tag{4.27}$$

By (4.23),

$$B_n(x+1) = \sum_{k=0}^{n}\binom{n}{k}B_k(x) = \sum_{k=0}^{n-1}\binom{n}{k}B_k(x)+B_n(x),$$

whence, by (4.24),

$$\sum_{k=0}^{n-1}\binom{n}{k}B_k(x) = n\,x^{n-1}. \tag{4.28}$$

Moreover, by (4.19), for $|z| < 2\pi$ we have

$$\begin{aligned}\sum_{n=0}^{\infty}\frac{z^n}{n!}\sum_{k=0}^{m-1}B_n\Big(x+\frac{k}{m}\Big) &= \sum_{k=0}^{m-1}\sum_{n=0}^{\infty}B_n\Big(x+\frac{k}{m}\Big)\frac{z^n}{n!} = \sum_{k=0}^{m-1}\frac{z\,e^{(x+k/m)z}}{e^z-1} \\ &= \frac{z\,e^{xz}}{e^z-1}\sum_{k=0}^{m-1}\big(e^{z/m}\big)^k = \frac{z\,e^{xz}}{e^z-1}\,\frac{e^z-1}{e^{z/m}-1} = m\,\frac{(z/m)\,e^{xz}}{e^{z/m}-1} \\ &= m\sum_{n=0}^{\infty}B_n(mx)\,\frac{(z/m)^n}{n!} = \sum_{n=0}^{\infty}\frac{z^n}{n!}\,\frac{B_n(mx)}{m^{n-1}}.\end{aligned}$$

Hence we get the multiplication formula for the Bernoulli polynomials:

$$B_n(mx) = m^{n-1}\sum_{k=0}^{m-1}B_n\Big(x+\frac{k}{m}\Big) \quad (n=0,1,2,\ldots;\ m=1,2,\ldots). \tag{4.29}$$

Chapter 5
Summation Formulae

5.1 Stirling's Formula for $n!$

For $n \in \mathbb{N}$ let

$$S_n := \int_0^{\pi/2} \sin^n x \, dx.$$

Integrating by parts we get, for $n \geq 2$,

$$S_n = -\Big[\sin^{n-1} x \, \cos x \Big]_0^{\pi/2} + (n-1) \int_0^{\pi/2} \cos^2 x \, \sin^{n-2} x \, dx$$

$$= (n-1) \int_0^{\pi/2} \big(1 - \sin^2 x\big) \sin^{n-2} x \, dx = (n-1) S_{n-2} - (n-1) S_n,$$

whence the recurrence formula

$$S_n = \frac{n-1}{n} S_{n-2} \qquad (n \geq 2). \tag{5.1}$$

Since $S_0 = \pi/2$ and $S_1 = 1$, by repeated application of (5.1) we obtain

$$S_{2n} = \frac{(2n-1)!!}{(2n)!!} \frac{\pi}{2}, \qquad S_{2n+1} = \frac{(2n)!!}{(2n+1)!!}, \tag{5.2}$$

C. Viola, *An Introduction to Special Functions*,
UNITEXT - La Matematica per il 3+2 102, DOI 10.1007/978-3-319-41345-7_5

with the usual notation

$$
\begin{aligned}
(2n-1)!! &= 1\cdot 3\cdot 5\cdots(2n-1),\\
(2n)!! &= 2\cdot 4\cdot 6\cdots(2n)
\end{aligned}
$$

for the semifactorial $N!!$ of a positive integer N.

For $0 \le x \le \pi/2$ we have $0 \le \sin x \le 1$, whence $\sin x \ge \sin^2 x \ge \sin^3 x \ge \ldots$. Therefore, integrating from 0 to $\pi/2$,

$$S_1 \ge S_2 \ge S_3 \ge \ldots .$$

Hence, by (5.2),

$$1 \ge \frac{S_{2n+1}}{S_{2n}} \ge \frac{S_{2n+1}}{S_{2n-1}} = \frac{2n}{2n+1} = 1 - \frac{1}{2n+1} = 1 + O\Big(\frac{1}{n}\Big).$$

Thus

$$\frac{S_{2n+1}}{S_{2n}} = 1 + O\Big(\frac{1}{n}\Big),$$

i.e., by (5.2),

$$\left(\frac{(2n)!!}{(2n-1)!!}\right)^2 \frac{1}{2n+1} = \prod_{k=1}^{n}\left(\frac{2k}{2k-1}\,\frac{2k}{2k+1}\right) = \frac{\pi}{2} + O\Big(\frac{1}{n}\Big). \tag{5.3}$$

For $n \to \infty$ we get Wallis' formula:

$$\frac{\pi}{2} = \prod_{k=1}^{\infty} \frac{4k^2}{4k^2-1} = \frac{2}{1}\,\frac{2}{3}\,\frac{4}{3}\,\frac{4}{5}\,\frac{6}{5}\,\frac{6}{7}\cdots .$$

Moreover, by (5.3),

$$\left(\frac{(2n)!!}{(2n-1)!!}\right)^2 = \frac{\pi}{2}(2n+1)\Big(1 + O\Big(\frac{1}{n}\Big)\Big) = \pi n\,\Big(1 + O\Big(\frac{1}{n}\Big)\Big). \tag{5.4}$$

By Taylor's formula, $(1+x)^{1/2} = 1 + \frac{1}{2}x + \binom{1/2}{2}x^2 + \ldots = 1 + O(x)\ \ (x \to 0)$. Hence

$$\Big(1 + O\Big(\frac{1}{n}\Big)\Big)^{1/2} = 1 + O\Big(\frac{1}{n}\Big) \quad (n \to \infty).$$

Also $(2n-1)!!\,(2n)!! = (2n)!$ and $(2n)!! = 2^n n!$, whence

$$\frac{(2n)!!}{(2n-1)!!} = \frac{(2n)!!^2}{(2n)!} = \frac{2^{2n}\,n!^2}{(2n)!}.$$

Thus from (5.4) we get

$$\frac{2^{2n}\, n!^2}{(2n)!} = \sqrt{\pi n}\left(1 + O\left(\frac{1}{n}\right)\right) \quad (n \to \infty). \tag{5.5}$$

We can now prove the following

Theorem 5.1 (Stirling's 'elementary' formula) *For $n \to \infty$ we have*

$$\log n! = \left(n + \frac{1}{2}\right)\log n - n + \log\sqrt{2\pi} + O\left(\frac{1}{n}\right) \tag{5.6}$$

and

$$n! = \sqrt{2\pi n}\left(\frac{n}{e}\right)^n\left(1 + O\left(\frac{1}{n}\right)\right). \tag{5.7}$$

Proof Plainly (5.6) and (5.7) are equivalent since, by Taylor's formula, $\exp O(\frac{1}{n}) = 1 + O(\frac{1}{n})$ and $\log(1 + O(\frac{1}{n})) = O(\frac{1}{n})$. We prove (5.6).

Let $k \geq 1$ be an integer. Then

$$\int_{k-1/2}^{k+1/2} \log x \,\mathrm{d}x = \int_0^{1/2} \big(\log(k-x) + \log(k+x)\big)\,\mathrm{d}x = \int_0^{1/2} \log(k^2 - x^2)\,\mathrm{d}x \tag{5.8}$$

$$= \int_0^{1/2}\left(\log k^2 + \log\left(1 - \frac{x^2}{k^2}\right)\right)\mathrm{d}x = \log k - R_k,$$

where

$$R_k = -\int_0^{1/2} \log\left(1 - \frac{x^2}{k^2}\right)\mathrm{d}x = O\left(\frac{1}{k^2}\right) \tag{5.9}$$

since $-\log\left(1 - x^2/k^2\right) = O\left(1/k^2\right)$ uniformly for $0 \leq x \leq 1/2$. Thus the series $\sum_{k=1}^{\infty} R_k$ converges. From (5.8) we get

$$\log n! = \sum_{k=1}^{n} \log k = \sum_{k=1}^{n} \int_{k-1/2}^{k+1/2} \log x \,\mathrm{d}x + \sum_{k=1}^{\infty} R_k - \sum_{k=n+1}^{\infty} R_k. \tag{5.10}$$

By (5.9)

$$\sum_{k=n+1}^{\infty} R_k = O\Big(\sum_{k=n+1}^{\infty} \frac{1}{k^2}\Big),$$

and clearly

$$\sum_{k=n+1}^{\infty} \frac{1}{k^2} < \int_n^{\infty} \frac{\mathrm{d}x}{x^2} = \frac{1}{n}.$$

Therefore, by (5.10),

$$\begin{aligned}
\log n! &= \int_{1/2}^{n+1/2} \log x \,\mathrm{d}x + C_1 + O\Big(\frac{1}{n}\Big) \\
&= \Big(n+\frac{1}{2}\Big)\log\Big(n+\frac{1}{2}\Big) - n + C_2 + O\Big(\frac{1}{n}\Big) \\
&= \Big(n+\frac{1}{2}\Big)\Big(\log n + \log\Big(1+\frac{1}{2n}\Big)\Big) - n + C_2 + O\Big(\frac{1}{n}\Big) \\
&= \Big(n+\frac{1}{2}\Big)\Big(\log n + \frac{1}{2n} + O\Big(\frac{1}{n^2}\Big)\Big) - n + C_2 + O\Big(\frac{1}{n}\Big),
\end{aligned}$$

for suitable constants C_1 and C_2. It follows that

$$\log n! = \Big(n+\frac{1}{2}\Big)\log n - n + C + O\Big(\frac{1}{n}\Big), \tag{5.11}$$

where C is a constant. To compute C we take the logarithm of (5.5). We get

$$2n\log 2 + 2\log n! - \log(2n)! = \log\sqrt{\pi} + (1/2)\log n + O(1/n).$$

Here we replace $\log n!$ and $\log(2n)!$ by the quantities obtained from the right-hand side of (5.11). This easily yields

$$C = (1/2)\log 2 + \log\sqrt{\pi} + O(1/n).$$

Since C is constant, for $n \to \infty$ we get

$$C = \log\sqrt{2\pi}. \tag{5.12}$$

Then (5.6) follows from (5.11) and (5.12). □

In Theorem 6.4 and Corollary 6.2 we shall generalize and improve (5.6) by proving Stirling's formula for the Euler gamma-function, with an asymptotic expansion of the remainder term.

From (4.17) and (5.7) we obtain an asymptotic formula for the Bernoulli numbers B_{2k}. Since

$$\sum_{n=2}^{\infty} n^{-2k} < \int_{1}^{\infty} x^{-2k}\, \mathrm{d}x = \frac{1}{2k-1} = O\Big(\frac{1}{k}\Big)$$

we get

$$1 < \zeta(2k) = 1 + \sum_{n=2}^{\infty} n^{-2k} < 1 + O\Big(\frac{1}{k}\Big),$$

whence

$$\zeta(2k) = 1 + O\Big(\frac{1}{k}\Big).$$

Thus, by (4.17) and (5.7),

$$|B_{2k}| = \frac{2}{(2\pi)^{2k}} \sqrt{4k\pi} \Big(\frac{2k}{e}\Big)^{2k} \Big(1 + O\Big(\frac{1}{k}\Big)\Big),$$

i.e.,

$$|B_{2k}| = 4\sqrt{k\pi} \Big(\frac{k}{e\pi}\Big)^{2k} \Big(1 + O\Big(\frac{1}{k}\Big)\Big). \tag{5.13}$$

5.2 Partial Summation and Euler's Constant

Our next task is to prove a partial summation formula, a variant of which was introduced by Abel in connection with the proof of Abel's theorem on uniform convergence of power series.

Theorem 5.2 (Abel's partial summation) *Let $\lambda_1 \le \lambda_2 \le \lambda_3 \le \dots$ be a sequence of real numbers such that* $\lim \lambda_n = \infty$. *Let $x > \lambda_1$ be a real number, and let $f(t)$ be a complex valued function, continuously differentiable in the real interval $\lambda_1 \le t \le x$. Let $a_1, a_2, a_3, \dots$ be an arbitrary sequence of complex numbers. Then*

$$\sum_{\substack{n \\ \lambda_n \le x}} a_n\, f(\lambda_n) = \Big(\sum_{\substack{n \\ \lambda_n \le x}} a_n\Big) f(x) - \int_{\lambda_1}^{x} \Big(\sum_{\substack{n \\ \lambda_n \le t}} a_n\Big) f'(t)\, \mathrm{d}t. \tag{5.14}$$

Proof Let N be such that $\lambda_N \le x < \lambda_{N+1}$. Then

$$\begin{aligned}
\Big(\sum_{\substack{n\\ \lambda_n\le x}} a_n\Big) f(x) - \sum_{\substack{n\\ \lambda_n\le x}} a_n\, f(\lambda_n) &= \sum_{n=1}^{N} a_n\big(f(x) - f(\lambda_n)\big)\\
&= \sum_{n=1}^{N} \int_{\lambda_n}^{x} a_n\, f'(t)\,\mathrm{d}t\\
&= \int_{\lambda_1}^{\lambda_2} a_1\, f'(t)\,\mathrm{d}t + \int_{\lambda_2}^{\lambda_3} a_1\, f'(t)\,\mathrm{d}t + \cdots + \int_{\lambda_N}^{x} a_1\, f'(t)\,\mathrm{d}t\\
&\qquad + \int_{\lambda_2}^{\lambda_3} a_2\, f'(t)\,\mathrm{d}t + \cdots + \int_{\lambda_N}^{x} a_2\, f'(t)\,\mathrm{d}t\\
&\qquad + \cdots\cdots\cdots\cdots\cdots\cdots\\
&\qquad + \int_{\lambda_N}^{x} a_N\, f'(t)\,\mathrm{d}t\\
&= \int_{\lambda_1}^{\lambda_2} a_1\, f'(t)\,\mathrm{d}t + \int_{\lambda_2}^{\lambda_3} (a_1 + a_2) f'(t)\,\mathrm{d}t + \cdots\\
&\qquad + \int_{\lambda_N}^{x} (a_1 + a_2 + \cdots + a_N) f'(t)\,\mathrm{d}t\\
&= \int_{\lambda_1}^{x} \Big(\sum_{\substack{n\\ \lambda_n\le t}} a_n\Big) f'(t)\,\mathrm{d}t.
\end{aligned}$$

□

The meaning of Theorem 5.2 is that, by introducing twice on the right-hand side of (5.14) suitable information on the partial sum $\sum a_n$, one gets information on the full sum $\sum a_n f(\lambda_n)$. We give an instance of this by proving an asymptotic formula for the sum $\sum_{n\le x} 1/n$, which also allows us to define Euler's constant γ.

We recall that $\lfloor x\rfloor$ denotes the integer part of the real number x, and $\{x\} = x - \lfloor x\rfloor$ denotes the fractional part of x. In Theorem 5.2 we choose $\lambda_n = n$, $a_n = 1$ ($n = 1, 2, 3, \ldots$), and $f(t) = 1/t$. By (5.14) we get

$$\sum_{n=1}^{\lfloor x\rfloor}\frac{1}{n}=\Big(\sum_{n=1}^{\lfloor x\rfloor}1\Big)\frac{1}{x}+\int_1^x\Big(\sum_{n=1}^{\lfloor t\rfloor}1\Big)\frac{\mathrm{d}t}{t^2}=\frac{\lfloor x\rfloor}{x}+\int_1^x\frac{\lfloor t\rfloor}{t^2}\,\mathrm{d}t$$

$$=1-\frac{\{x\}}{x}+\int_1^x\Big(\frac{1}{t}-\frac{\{t\}}{t^2}\Big)\mathrm{d}t.$$

Since $\{t\}/t^2<1/t^2$, the integral $\int_1^\infty(\{t\}/t^2)\,\mathrm{d}t$ converges. Hence

$$\sum_{n=1}^{\lfloor x\rfloor}\frac{1}{n}=\log x-\frac{\{x\}}{x}+\gamma+\int_x^\infty\frac{\{t\}}{t^2}\,\mathrm{d}t, \tag{5.15}$$

where $\gamma=1-\int_1^\infty(\{t\}/t^2)\,\mathrm{d}t$ is Euler's constant. For $x>0$ we have $\{x\}/x<1/x$ and $\int_x^\infty(\{t\}/t^2)\,\mathrm{d}t<\int_x^\infty \mathrm{d}t/t^2=1/x$. By (5.15) we obtain the asymptotic formula

$$\sum_{n=1}^{\lfloor x\rfloor}\frac{1}{n}=\log x+\gamma+O\Big(\frac{1}{x}\Big)\qquad(x\to+\infty). \tag{5.16}$$

Since $\log x=\log\lfloor x\rfloor+\log(1+\{x\}/\lfloor x\rfloor)=\log\lfloor x\rfloor+O(1/\lfloor x\rfloor)$, writing $\lfloor x\rfloor=N$ we get from (5.16)

$$\sum_{n=1}^{N}\frac{1}{n}=\log N+\gamma+O\Big(\frac{1}{N}\Big)\qquad(N\to+\infty), \tag{5.17}$$

whence

$$\gamma=1-\int_1^\infty\frac{\{t\}}{t^2}\,\mathrm{d}t=\lim_{N\to\infty}\Big(\sum_{n=1}^{N}\frac{1}{n}-\log N\Big)=0.577215\ldots. \tag{5.18}$$

We shall prove two further integral representations ((5.23) and (5.24)) of Euler's constant γ. First of all we remark that, for any positive integer N and for $0<t<N$,

$$1+\frac{t}{N}<1+\frac{t}{N}+\frac{t^2}{2!\,N^2}+\ldots<1+\frac{t}{N}+\frac{t^2}{N^2}+\ldots,$$

i.e.,

$$1+\frac{t}{N}<e^{t/N}<\Big(1-\frac{t}{N}\Big)^{-1},$$

whence $(1+t/N)^N < e^t$ and $(1-t/N)^N < e^{-t}$. Therefore

$$0 < e^{-t} - \left(1 - \frac{t}{N}\right)^N = e^{-t}\left(1 - e^t\left(1 - \frac{t}{N}\right)^N\right) \tag{5.19}$$
$$< e^{-t}\left(1 - \left(1 - \frac{t^2}{N^2}\right)^N\right).$$

Comparing the graphs of the functions $y = (1-x)^N$ and $y = 1 - Nx$ we see that $1 - Nx < (1-x)^N$ for all $0 < x \leq 1/N$. Substituting $x = t^2/N^2$ we get

$$1 - \left(1 - \frac{t^2}{N^2}\right)^N < \frac{t^2}{N} \quad \text{for } 0 < t \leq \sqrt{N}, \tag{5.20}$$

and (5.20) is trivially true for $\sqrt{N} < t < N$. Hence, by (5.19) and (5.20),

$$0 < e^{-t} - \left(1 - \frac{t}{N}\right)^N < \frac{t^2}{N} e^{-t} \quad \text{for } 0 < t < N. \tag{5.21}$$

With the substitution $1 - t/N = u$ we get

$$\int_0^N \left(1 - \left(1 - \frac{t}{N}\right)^N\right)\frac{\mathrm{d}t}{t} = \int_0^1 \frac{1-u^N}{1-u}\, du = \int_0^1 (1 + u + u^2 + \cdots + u^{N-1})\, du$$
$$= 1 + \frac{1}{2} + \frac{1}{3} + \cdots + \frac{1}{N}.$$

Thus

$$\sum_{n=1}^N \frac{1}{n} = \int_0^N \left(e^{-t} - \left(1 - \frac{t}{N}\right)^N\right)\frac{\mathrm{d}t}{t} + \int_0^N \frac{1-e^{-t}}{t}\,\mathrm{d}t$$
$$= \int_0^N \left(e^{-t} - \left(1 - \frac{t}{N}\right)^N\right)\frac{\mathrm{d}t}{t} + \int_0^1 \frac{1-e^{-t}}{t}\,\mathrm{d}t + \int_1^N \frac{\mathrm{d}t}{t} - \int_1^N \frac{e^{-t}}{t}\,\mathrm{d}t.$$

It follows that

$$\sum_{n=1}^N \frac{1}{n} - \log N = \int_0^N \left(e^{-t} - \left(1 - \frac{t}{N}\right)^N\right)\frac{\mathrm{d}t}{t} + \int_0^1 \frac{1-e^{-t}}{t}\,\mathrm{d}t - \int_1^N \frac{e^{-t}}{t}\,\mathrm{d}t. \tag{5.22}$$

By (5.21),

$$0 < \int_0^N \left(e^{-t} - \left(1 - \frac{t}{N}\right)^N \right) \frac{dt}{t} < \frac{1}{N} \int_0^N t e^{-t}\, dt$$

$$< \frac{1}{N} \int_0^\infty t e^{-t}\, dt \to 0 \quad (N \to \infty).$$

Hence by (5.18) and (5.22) we get, for $N \to \infty$,

$$\gamma = \int_0^1 \frac{1 - e^{-t}}{t}\, dt - \int_1^\infty \frac{e^{-t}}{t}\, dt = \int_0^1 \frac{1 - e^{-t} - e^{-1/t}}{t}\, dt. \tag{5.23}$$

From (5.23) we obtain

$$\gamma = \lim_{\delta \to 0+} \left(\int_\delta^1 \frac{dt}{t} - \int_\delta^1 \frac{e^{-t}}{t}\, dt - \int_1^\infty \frac{e^{-t}}{t}\, dt \right) = \lim_{\delta \to 0+} \left(\int_\delta^1 \frac{dt}{t} - \int_\delta^\infty \frac{e^{-t}}{t}\, dt \right).$$

We have

$$\int_\delta^1 \frac{dt}{t} = \int_{1-e^{-\delta}}^1 \frac{dt}{t} - \int_{1-e^{-\delta}}^\delta \frac{dt}{t} = \int_{1-e^{-\delta}}^1 \frac{dt}{t} - \log \frac{\delta}{1 - e^{-\delta}},$$

and, with the substitution $t = 1 - e^{-u}$,

$$\int_{1-e^{-\delta}}^1 \frac{dt}{t} = \int_\delta^\infty \frac{e^{-u}}{1 - e^{-u}}\, du.$$

Therefore

$$\gamma = \lim_{\delta \to 0+} \left(\int_\delta^\infty \frac{e^{-t}}{1 - e^{-t}}\, dt - \int_\delta^\infty \frac{e^{-t}}{t}\, dt \right) - \lim_{\delta \to 0+} \log \frac{\delta}{1 - e^{-\delta}},$$

whence

$$\gamma = \int_0^\infty \left(\frac{1}{1 - e^{-t}} - \frac{1}{t} \right) e^{-t}\, dt. \tag{5.24}$$

5.3 The Euler–MacLaurin Summation Formula

Theorem 5.3 (Euler–MacLaurin's summation formula) *Let $q \geq 1$ be an integer, and let $f(t)$ be a complex valued function having continuous qth derivative in a real interval $a \leq t \leq b$. Then*

$$\sum_{a<n\leq b} f(n) = \int_a^b f(t)\,\mathrm{d}t + \sum_{k=1}^{q} \frac{(-1)^k}{k!}\Big[B_k(\{t\})\, f^{(k-1)}(t)\Big]_a^b + \frac{(-1)^{q+1}}{q!}\int_a^b B_q(\{t\})\, f^{(q)}(t)\,\mathrm{d}t, \tag{5.25}$$

where n is integer and $B_k(\{t\})$ is the value of the kth Bernoulli polynomial at the fractional part $\{t\}$ of t. In particular, choosing $a = M$ and $b = N$ with M, N integers, we obtain

$$\sum_{n=M}^{N} f(n) = \int_M^N f(t)\,\mathrm{d}t + \frac{f(M)+f(N)}{2} + \sum_{h=1}^{\lfloor q/2\rfloor} \frac{B_{2h}}{(2h)!}\big(f^{(2h-1)}(N) - f^{(2h-1)}(M)\big) + \frac{(-1)^{q+1}}{q!}\int_M^N B_q(\{t\})\, f^{(q)}(t)\,\mathrm{d}t, \tag{5.26}$$

where B_{2h} is the $(2h)$th Bernoulli number.

Proof Since $B_1(t) = t - 1/2$, for $q = 1$ and for n integer we get, integrating by parts,

$$\int_n^{n+1} B_1(\{t\})\, f'(t)\,\mathrm{d}t = \int_n^{n+1}\Big(t - n - \frac{1}{2}\Big) f'(t)\,\mathrm{d}t = \frac{f(n)+f(n+1)}{2} - \int_n^{n+1} f(t)\,\mathrm{d}t,$$

and

$$\int_a^{\lfloor a\rfloor+1} B_1(\{t\})\, f'(t)\,\mathrm{d}t = \int_a^{\lfloor a\rfloor+1}\Big(t - \lfloor a\rfloor - \frac{1}{2}\Big) f'(t)\,\mathrm{d}t$$

$$= \frac{f(\lfloor a\rfloor+1)}{2} - \Big(\{a\} - \frac{1}{2}\Big) f(a) - \int_a^{\lfloor a\rfloor+1} f(t)\,\mathrm{d}t,$$

$$\int_{\lfloor b\rfloor}^{b} B_1(\{t\})\, f'(t)\,\mathrm{d}t = \int_{\lfloor b\rfloor}^{b} \left(t - \lfloor b\rfloor - \frac{1}{2}\right) f'(t)\,\mathrm{d}t$$

$$= \frac{f(\lfloor b\rfloor)}{2} + \left(\{b\} - \frac{1}{2}\right) f(b) - \int_{\lfloor b\rfloor}^{b} f(t)\,\mathrm{d}t.$$

Therefore

$$\int_{a}^{b} B_1(\{t\})\, f'(t)\,\mathrm{d}t \tag{5.27}$$

$$= \int_{a}^{\lfloor a\rfloor+1} B_1(\{t\})\, f'(t)\,\mathrm{d}t + \sum_{n=\lfloor a\rfloor+1}^{\lfloor b\rfloor-1} \int_{n}^{n+1} B_1(\{t\})\, f'(t)\,\mathrm{d}t + \int_{\lfloor b\rfloor}^{b} B_1(\{t\})\, f'(t)\,\mathrm{d}t$$

$$= \sum_{n=\lfloor a\rfloor+1}^{\lfloor b\rfloor} f(n) - \int_{a}^{b} f(t)\,\mathrm{d}t + \Big[B_1(\{t\})\, f(t)\Big]_a^b,$$

which proves (5.25) for $q = 1$.

For $q \geq 2$ we have, by (4.21) and (4.22), $B_q(0) = B_q(1) = B_q$. Hence the periodic function $B_q(\{t\})$ is continuous. Consequently, integrating by parts,

$$\frac{(-1)^{q+1}}{q!} \int_{a}^{b} B_q(\{t\})\, f^{(q)}(t)\,\mathrm{d}t$$

$$= \frac{(-1)^{q+1}}{q!} \Big[B_q(\{t\})\, f^{(q-1)}(t)\Big]_a^b - \frac{(-1)^{q+1}}{q!} \int_{a}^{b} B_q'(\{t\})\, f^{(q-1)}(t)\,\mathrm{d}t,$$

whence, by (4.25),

$$\frac{(-1)^{q+1}}{q!} \int_{a}^{b} B_q(\{t\})\, f^{(q)}(t)\,\mathrm{d}t \tag{5.28}$$

$$= -\frac{(-1)^{q}}{q!} \Big[B_q(\{t\})\, f^{(q-1)}(t)\Big]_a^b + \frac{(-1)^{q}}{(q-1)!} \int_{a}^{b} B_{q-1}(\{t\})\, f^{(q-1)}(t)\,\mathrm{d}t.$$

By repeated application of (5.28) and by (5.27) we have, for $q \geq 2$,

$$\frac{(-1)^{q+1}}{q!} \int_a^b B_q(\{t\})\, f^{(q)}(t)\,\mathrm{d}t$$

$$= -\sum_{k=2}^{q} \frac{(-1)^k}{k!} \Big[B_k(\{t\})\, f^{(k-1)}(t) \Big]_a^b + \int_a^b B_1(\{t\})\, f'(t)\,\mathrm{d}t$$

$$= \sum_{a<n\leq b} f(n) - \int_a^b f(t)\,\mathrm{d}t - \sum_{k=1}^{q} \frac{(-1)^k}{k!} \Big[B_k(\{t\})\, f^{(k-1)}(t) \Big]_a^b,$$

whence (5.25).

For integers $M < N$ we apply (5.25) with $a = M$ and $b = N$. Since $B_k(\{M\}) = B_k(\{N\}) = B_k(0) = B_k$, we obtain

$$\sum_{n=M+1}^{N} f(n) = \int_M^N f(t)\,\mathrm{d}t + \sum_{k=1}^{q} (-1)^k \frac{B_k}{k!} \big(f^{(k-1)}(N) - f^{(k-1)}(M) \big) + \frac{(-1)^{q+1}}{q!} \int_M^N B_q(\{t\})\, f^{(q)}(t)\,\mathrm{d}t.$$

Adding $f(M)$, and recalling that $B_1 = -1/2$ and $B_{2h+1} = 0$ for $h \geq 1$, we have

$$\sum_{n=M}^{N} f(n) = \int_M^N f(t)\,\mathrm{d}t + f(M) + \frac{1}{2}\big(f(N) - f(M) \big) + \sum_{\substack{k=2 \\ k \text{ even}}}^{q} \frac{B_k}{k!} \big(f^{(k-1)}(N) - f^{(k-1)}(M) \big) + \frac{(-1)^{q+1}}{q!} \int_M^N B_q(\{t\})\, f^{(q)}(t)\,\mathrm{d}t.$$

Writing in the last sum $k = 2h$, we get (5.26). □

We compare the asymptotic formulae for the sum $\sum_{n=1}^{N} 1/n$ obtained by applying Theorems 5.2 and 5.3. In (5.26) we choose $M = 1$, $f(t) = 1/t$ and $q = 2r + 1$. For any integer $r \geq 0$ we obtain

$$\sum_{n=1}^{N} \frac{1}{n} = \log N + \frac{1}{2} + \frac{1}{2N} + \sum_{h=1}^{r} \frac{B_{2h}}{2h} \Big(1 - \frac{1}{N^{2h}} \Big) - \int_1^N \frac{B_{2r+1}(\{t\})}{t^{2r+2}}\,\mathrm{d}t.$$

Since the function $B_{2r+1}(\{t\})$ is bounded, the integral $\int_1^\infty (B_{2r+1}(\{t\})/t^{2r+2})\,\mathrm{d}t$ converges. Thus we get the following expansion of the remainder term in (5.17):

$$\sum_{n=1}^{N} \frac{1}{n} = \log N + \gamma + \frac{1}{2N} - \sum_{h=1}^{r} \frac{B_{2h}}{2h\,N^{2h}} + \int_N^\infty \frac{B_{2r+1}(\{t\})}{t^{2r+2}}\,\mathrm{d}t, \qquad (5.29)$$

where

$$\gamma = \frac{1}{2} + \sum_{h=1}^{r} \frac{B_{2h}}{2h} - \int_1^\infty \frac{B_{2r+1}(\{t\})}{t^{2r+2}}\,\mathrm{d}t = \lim_{N\to\infty} \left(\sum_{n=1}^{N} \frac{1}{n} - \log N \right)$$

is Euler's constant.

Plainly

$$\int_N^\infty \frac{B_{2r+1}(\{t\})}{t^{2r+2}}\,\mathrm{d}t \ll \int_N^\infty \frac{\mathrm{d}t}{t^{2r+2}} = \frac{1}{(2r+1)N^{2r+1}}.$$

Therefore, changing r to $r+1$ in (5.29) we obtain, for any integer $r \geq 0$, the following improvement upon (5.17):

$$\sum_{n=1}^{N} \frac{1}{n} = \log N + \gamma + \frac{1}{2N} - \sum_{h=1}^{r} \frac{B_{2h}}{2h\,N^{2h}} + O_r\Big(\frac{1}{N^{2r+2}}\Big) \quad (N \to +\infty). \quad (5.30)$$

This formula can also symbolically be written as an asymptotic series expansion:

$$\sum_{n=1}^{N} \frac{1}{n} \sim \log N + \gamma + \frac{1}{2N} - \sum_{h=1}^{\infty} \frac{B_{2h}}{2h\,N^{2h}} \qquad (N \to +\infty), \qquad (5.31)$$

where the series diverges for every N since the power series $\sum_{h=1}^{\infty}(B_{2h}/2h)x^h$ has radius of convergence zero, an easy consequence of the asymptotic formula (5.13) for $|B_{2h}|$. As appears in (5.30), the symbol $\sim$ in the asymptotic expansion (5.31) means that it is to be replaced by the symbol $=$ provided the divergent series in (5.31) is replaced by any partial sum $\sum_{h=1}^{r} B_{2h}/(2h\,N^{2h})$ plus an error term having the order of magnitude of the first subsequent term in the series.

5.4 Kronecker's Summation Formula

We shall also require Kronecker's summation formula (5.34) for a function of complex variable with controlled growth on a vertical strip.

Theorem 5.4 (Kronecker) *Let* $r, s \in \mathbb{Z}$, $r < s$, *and let* $f(z)$ *be regular in the strip* $r \le \operatorname{Re} z \le s$ *and satisfying*

$$|f(z)| = o\left(e^{2\pi|\operatorname{Im} z|}\right) \qquad (|\operatorname{Im} z| \to \infty) \tag{5.32}$$

and

$$\int_0^{+\infty} \frac{|f(x \pm iy)|}{e^{2\pi y}}\, \mathrm{d}y < +\infty \qquad (x \in \mathbb{Z},\ r \le x \le s). \tag{5.33}$$

Then

$$\frac{1}{2} f(r) + \sum_{k=r+1}^{s-1} f(k) + \frac{1}{2} f(s) \tag{5.34}$$

$$= \int_r^s f(x)\, \mathrm{d}x + i \int_0^{+\infty} \frac{f(r+iy) - f(r-iy)}{e^{2\pi y} - 1}\, \mathrm{d}y - i \int_0^{+\infty} \frac{f(s+iy) - f(s-iy)}{e^{2\pi y} - 1}\, \mathrm{d}y.$$

Proof By the assumption (5.33) and by the Taylor expansions of $f(z)$ at $z = r$ and $z = s$, the integrals on the right-hand side of (5.34) are absolutely convergent.

Let k be an integer such that $r \le k < s$. For $0 < \delta < 1/2 < \varDelta$, in the upper half-strip $\{k \le \operatorname{Re} z \le k+1,\ \operatorname{Im} z \ge 0\}$ we consider the contour $\lambda_{\delta,\varDelta} = \lambda_1 \cup \cdots \cup \lambda_6$ taken in the negative (i.e., clockwise) sense, where

$$\begin{aligned}
\lambda_1 &= \{k+1+iy \mid \delta \le y \le \varDelta\}, & \lambda_2 &= \{k+1+\delta e^{i\vartheta} \mid \pi/2 \le \vartheta \le \pi\},\\
\lambda_3 &= \{x \mid k+\delta \le x \le k+1-\delta\}, & \lambda_4 &= \{k+\delta e^{i\vartheta} \mid 0 \le \vartheta \le \pi/2\},\\
\lambda_5 &= \{k+iy \mid \delta \le y \le \varDelta\}, & \lambda_6 &= \{x+i\varDelta \mid k \le x \le k+1\}.
\end{aligned}$$

Since the function $f(z)/(e^{-2\pi i z} - 1)$ is regular in the strip $r \le \operatorname{Re} z \le s$ except at points $z \in \mathbb{Z}$, and $\lambda_{\delta,\varDelta}$ is the border of a rectangle indented at the corners $z = k$ and $z = k+1$, by Cauchy's theorem we get

$$\int_{\lambda_{\delta,\varDelta}} \frac{f(z)\, \mathrm{d}z}{e^{-2\pi i z} - 1} = 0.$$

Also, for $z \in \lambda_6 = \{x + i\varDelta \mid k \le x \le k+1\}$ we have

$$\left| \frac{f(z)}{e^{-2\pi i z} - 1} \right| \le \frac{|f(z)|}{e^{2\pi\varDelta} - 1}.$$

Hence, by (5.32),

$$\lim_{\Delta\to+\infty}\int_{\lambda_6}\frac{f(z)\,\mathrm{d}z}{e^{-2\pi iz}-1}=0.$$

Therefore

$$i\int_{\delta}^{+\infty}\frac{f(k+1+iy)}{e^{2\pi y}-1}\,\mathrm{d}y-i\int_{\delta}^{+\infty}\frac{f(k+iy)}{e^{2\pi y}-1}\,\mathrm{d}y+\int_{k+\delta}^{k+1-\delta}\frac{f(x)\,\mathrm{d}x}{e^{-2\pi ix}-1}\tag{5.35}$$
$$=\int_{\lambda_2}\frac{f(z)\,\mathrm{d}z}{e^{-2\pi iz}-1}+\int_{\lambda_4}\frac{f(z)\,\mathrm{d}z}{e^{-2\pi iz}-1}.$$

Symmetrically, in the lower half-strip $\{k\le \operatorname{Re} z\le k+1,\ \operatorname{Im} z\le 0\}$ we consider the integral

$$\int_{\overline{\lambda_{\delta,\Delta}}}\frac{f(z)\,\mathrm{d}z}{e^{2\pi iz}-1}$$

on the contour $\overline{\lambda_{\delta,\Delta}}$, conjugate to $\lambda_{\delta,\Delta}$, taken in the positive sense. We get

$$-i\int_{\delta}^{+\infty}\frac{f(k+1-iy)}{e^{2\pi y}-1}\,\mathrm{d}y+i\int_{\delta}^{+\infty}\frac{f(k-iy)}{e^{2\pi y}-1}\,\mathrm{d}y+\int_{k+\delta}^{k+1-\delta}\frac{f(x)\,\mathrm{d}x}{e^{2\pi ix}-1}\tag{5.36}$$
$$=\int_{\overline{\lambda_2}}\frac{f(z)\,\mathrm{d}z}{e^{2\pi iz}-1}+\int_{\overline{\lambda_4}}\frac{f(z)\,\mathrm{d}z}{e^{2\pi iz}-1}.$$

Since

$$\frac{1}{e^{-2\pi ix}-1}+\frac{1}{e^{2\pi ix}-1}=\frac{e^{2\pi ix}}{1-e^{2\pi ix}}-\frac{1}{1-e^{2\pi ix}}=-1,$$

summing (5.35) and (5.36) we obtain

$$-i\int_{\delta}^{+\infty}\frac{f(k+iy)-f(k-iy)}{e^{2\pi y}-1}\,\mathrm{d}y+i\int_{\delta}^{+\infty}\frac{f(k+1+iy)-f(k+1-iy)}{e^{2\pi y}-1}\,\mathrm{d}y$$
$$-\int_{k+\delta}^{k+1-\delta}f(x)\,\mathrm{d}x$$
$$=\int_{\lambda_2}\frac{f(z)\,\mathrm{d}z}{e^{-2\pi iz}-1}+\int_{\overline{\lambda_2}}\frac{f(z)\,\mathrm{d}z}{e^{2\pi iz}-1}+\int_{\lambda_4}\frac{f(z)\,\mathrm{d}z}{e^{-2\pi iz}-1}+\int_{\overline{\lambda_4}}\frac{f(z)\,\mathrm{d}z}{e^{2\pi iz}-1}.$$

For $\delta \to 0$ we get

$$\int_k^{k+1} f(x)\,\mathrm{d}x + i\int_0^{+\infty} \frac{f(k+iy)-f(k-iy)}{e^{2\pi y}-1}\,\mathrm{d}y \qquad (5.37)$$

$$- i\int_0^{+\infty} \frac{f(k+1+iy)-f(k+1-iy)}{e^{2\pi y}-1}\,\mathrm{d}y$$

$$= -\lim_{\delta\to 0}\left(\int_{\lambda_2} \frac{f(z)\,\mathrm{d}z}{e^{-2\pi iz}-1} + \int_{\overline{\lambda_2}} \frac{f(z)\,\mathrm{d}z}{e^{2\pi iz}-1}\right) - \lim_{\delta\to 0}\left(\int_{\lambda_4} \frac{f(z)\,\mathrm{d}z}{e^{-2\pi iz}-1} + \int_{\overline{\lambda_4}} \frac{f(z)\,\mathrm{d}z}{e^{2\pi iz}-1}\right).$$

We have

$$\int_{\lambda_4} \frac{f(z)\,\mathrm{d}z}{e^{-2\pi iz}-1} = f(k)\int_{\lambda_4} \frac{\mathrm{d}z}{e^{-2\pi iz}-1} + \int_{\lambda_4} \frac{f(z)-f(k)}{e^{-2\pi iz}-1}\,\mathrm{d}z, \qquad (5.38)$$

and

$$\lim_{z\to k} \frac{f(z)-f(k)}{e^{-2\pi iz}-1} = \lim_{z\to k} \frac{f'(z)}{-2\pi i\, e^{-2\pi iz}} = -\frac{f'(k)}{2\pi i}.$$

Therefore

$$\left|\int_{\lambda_4} \frac{f(z)-f(k)}{e^{-2\pi iz}-1}\,\mathrm{d}z\right| \le \int_{\lambda_4} \left|\frac{f(z)-f(k)}{e^{-2\pi iz}-1}\right| |\mathrm{d}z| \ll \int_{\lambda_4} |\mathrm{d}z| = \frac{\pi}{2}\,\delta,$$

whence, by (5.38),

$$\lim_{\delta\to 0}\int_{\lambda_4} \frac{f(z)\,\mathrm{d}z}{e^{-2\pi iz}-1} = f(k)\lim_{\delta\to 0}\int_{\lambda_4} \frac{\mathrm{d}z}{e^{-2\pi iz}-1}.$$

By a similar treatment of the other three integrals on the right-hand side of (5.37) we obtain

$$\int_k^{k+1} f(x)\,\mathrm{d}x + i\int_0^{+\infty} \frac{f(k+iy)-f(k-iy)}{e^{2\pi y}-1}\,\mathrm{d}y \qquad (5.39)$$

$$- i\int_0^{+\infty} \frac{f(k+1+iy)-f(k+1-iy)}{e^{2\pi y}-1}\,\mathrm{d}y$$

$$= -f(k+1)\lim_{\delta\to 0}\left(\int\limits_{\lambda_2}\frac{dz}{e^{-2\pi iz}-1}+\int\limits_{\overline{\lambda_2}}\frac{dz}{e^{2\pi iz}-1}\right)$$

$$- f(k)\lim_{\delta\to 0}\left(\int\limits_{\lambda_4}\frac{dz}{e^{-2\pi iz}-1}+\int\limits_{\overline{\lambda_4}}\frac{dz}{e^{2\pi iz}-1}\right).$$

Since the function $1/(e^{-2\pi iz}-1)$ has period 1, we have

$$\int\limits_{\lambda_4}\frac{dz}{e^{-2\pi iz}-1} = \int\limits_{\substack{|z|=\delta\\ 0\le\arg z\le\pi/2}}\frac{dz}{e^{-2\pi iz}-1}.$$

Moreover, for $z\to 0$,

$$e^{-2\pi iz}-1 = -2\pi iz-\frac{(2\pi)^2}{2!}z^2+\ldots = -2\pi iz\big(1+O(|z|)\big),$$

whence

$$\frac{1}{e^{-2\pi iz}-1} = -\frac{1}{2\pi iz}\,\frac{1}{1+O(|z|)} = -\frac{1}{2\pi iz}\big(1+O(|z|)\big).$$

Thus

$$\int\limits_{\substack{|z|=\delta\\ 0\le\arg z\le\pi/2}}\frac{dz}{e^{-2\pi iz}-1} = -\frac{1}{2\pi i}\int\limits_{\substack{|z|=\delta\\ 0\le\arg z\le\pi/2}}\frac{dz}{z}+O\Bigg(\int\limits_{|z|=\delta}|dz|\Bigg)$$

$$= -\frac{1}{2\pi i}\,i\int\limits_0^{\pi/2}d\vartheta+O(\delta) = -\frac{1}{4}+O(\delta),$$

whence

$$\lim_{\delta\to 0}\int\limits_{\lambda_4}\frac{dz}{e^{-2\pi iz}-1} = -\frac{1}{4}.$$

Similarly

$$\lim_{\delta\to 0}\int\limits_{\overline{\lambda_4}}\frac{dz}{e^{2\pi iz}-1} = \lim_{\delta\to 0}\int\limits_{\lambda_2}\frac{dz}{e^{-2\pi iz}-1} = \lim_{\delta\to 0}\int\limits_{\overline{\lambda_2}}\frac{dz}{e^{2\pi iz}-1} = -\frac{1}{4}.$$

From (5.39) we conclude that

$$\frac{1}{2}\,f(k)+\frac{1}{2}\,f(k+1)=\int\limits_{k}^{k+1} f(x)\,\mathrm{d}x$$

$$+\,i\int\limits_{0}^{+\infty}\frac{f(k+iy)-f(k-iy)}{e^{2\pi y}-1}\,\mathrm{d}y-i\int\limits_{0}^{+\infty}\frac{f(k+1+iy)-f(k+1-iy)}{e^{2\pi y}-1}\,\mathrm{d}y.$$

Summing this formula for $k=r, r+1, \dots, s-1$ we get (5.34). □

Chapter 6
The Euler Gamma-Function

6.1 Eulerian Integrals

The gamma-function $\Gamma(z)$ was introduced by Euler with the purpose of interpolating in a natural way the sequence $n!$ (see (6.9)). Euler defined $\Gamma(z)$ by means of the infinite product (6.18), and showed that $\Gamma(z)$ can be represented by the integral (6.1) if $z > 0$, or by the limit formula (6.14). Weierstrass defined $(z\Gamma(z))^{-1}$ through the infinite product (6.29). We prefer to use (6.1) as a definition.

Definition 6.1

$$\Gamma(z) = \int_0^{+\infty} e^{-t}\, t^{z-1}\, dt \qquad (z \in \mathbb{C},\ \operatorname{Re} z > 0). \tag{6.1}$$

The notation $\Gamma(z)$ is due to Legendre, who called the integral (6.1) eulerian integral of the second kind.

We recall that, for $\alpha, \beta \in \mathbb{C}$, $\alpha \neq 0$, the power α^β is defined by

$$\alpha^\beta = \exp(\beta \log \alpha) = \exp\big(\beta(\log|\alpha| + i \arg \alpha)\big), \tag{6.2}$$

so that, in general, α^β has infinitely many values corresponding to the infinitely many values of $\arg \alpha$ in (6.2) (more precisely, α^β has infinitely many distinct values if $\beta \in \mathbb{C} \setminus \mathbb{Q}$, or n distinct values if $\beta = m/n$ with $m, n \in \mathbb{Z}$, $n > 0$ and $\gcd(m, n) = 1$). However, if $\alpha \in \mathbb{R}$, $\alpha > 0$, unless otherwise specified one conventionally takes $\arg \alpha = 0$ in (6.2), so that $\alpha^\beta = \exp(\beta \log \alpha)$ with the elementary real value of $\log \alpha$. In particular this yields

$$\begin{aligned} |\alpha^\beta| &= |\exp(\beta \log \alpha)| = \exp \operatorname{Re}(\beta \log \alpha) \\ &= \exp\big((\operatorname{Re}\beta) \log \alpha\big) = \alpha^{\operatorname{Re}\beta} \qquad (\alpha > 0). \end{aligned} \tag{6.3}$$

C. Viola, *An Introduction to Special Functions*,
UNITEXT - La Matematica per il 3+2 102, DOI 10.1007/978-3-319-41345-7_6

This applies to $t^{z-1} = \exp\big((z-1)\log t\big)$ in (6.1), where $t \in (0, +\infty)$, as well as to the power $n^{-s} = \exp(-s \log n)$ $(n = 1, 2, \dots)$ in the definition (4.3) of the Riemann zeta-function.

By (6.3) we have $|e^{-t}\, t^{z-1}| = e^{-t}\, t^{\operatorname{Re} z-1}$, whence for any $z \in \mathbb{C}$ with $\operatorname{Re} z > 0$ the integral (6.1) converges absolutely both at $t = 0$ and at $t = +\infty$. Moreover

$$\int_0^{+\infty} |e^{-t}\, t^{z-1}|\, \mathrm{d}t = \int_0^1 e^{-t}\, t^{\operatorname{Re} z-1}\, \mathrm{d}t + \int_1^{+\infty} e^{-t}\, t^{\operatorname{Re} z-1}\, \mathrm{d}t. \tag{6.4}$$

For z in any vertical strip $a \le \operatorname{Re} z \le b$ with $a > 0$ and $b < +\infty$, we have $e^{-t} t^{\operatorname{Re} z-1} \le t^{a-1}$ in the first integral on the right-hand side of (6.4), and $e^{-t} t^{\operatorname{Re} z-1} \le e^{-t} t^{b-1}$ in the second integral. Since

$$\int_0^1 t^{a-1}\, \mathrm{d}t < +\infty \text{ for } a > 0 \text{ and } \int_1^{+\infty} e^{-t}\, t^{b-1}\, \mathrm{d}t < +\infty,$$

the decomposition (6.4) shows that the integral (6.1) converges uniformly in any strip $0 < a \le \operatorname{Re} z \le b$ and hence is a regular function of z in the strip. Since $a > 0$ can be taken arbitrarily small, and b arbitrarily large, $\Gamma(z)$ is regular in the open half-plane $\operatorname{Re} z > 0$.

For $\operatorname{Re} z > 0$ we have, by (6.3),

$$|t^z| = t^{\operatorname{Re} z} \to 0 \qquad (t \to 0+).$$

Thus from (6.1) we get, integrating by parts,

$$\Gamma(z) = \frac{1}{z}\Big[e^{-t}\, t^z\Big]_{t=0}^{+\infty} + \frac{1}{z} \int_0^{+\infty} e^{-t}\, t^z\, \mathrm{d}t = \frac{1}{z}\, \Gamma(z+1), \tag{6.5}$$

whence the first functional equation for the gamma-function:

$$\Gamma(z+1) = z\, \Gamma(z). \tag{6.6}$$

Since $\Gamma(z+1)$ is regular for $\operatorname{Re} z > -1$, (6.5) yields the analytic continuation of $\Gamma(z)$ in the half-plane $\operatorname{Re} z > -1$ and shows that $\Gamma(z)$ has a simple pole at $z = 0$ with residue

$$\Gamma(1) = \int_0^{+\infty} e^{-t}\, \mathrm{d}t = 1.$$

Plainly this argument can be iterated. By n-fold application of (6.5) we get

$$\Gamma(z) = \frac{1}{z}\,\Gamma(z+1) = \frac{1}{z(z+1)}\,\Gamma(z+2) = \cdots \tag{6.7}$$

$$= \frac{1}{z(z+1)\cdots(z+n-1)}\,\Gamma(z+n),$$

which yields the analytic continuation of $\Gamma(z)$ for $\operatorname{Re} z > -n$. Since n is arbitrary, we conclude that $\Gamma(z)$ is analytically continued in the whole $\mathbb{C}$ to a meromorphic function, regular in $\mathbb{C} \setminus \{0, -1, -2, \dots\}$ and having simple poles at $z = 0, -1, -2, \dots$. Moreover, changing n to $n+1$ in (6.7),

$$\Gamma(z) = \frac{1}{z+n} \cdot \frac{\Gamma(z+n+1)}{z(z+1)\cdots(z+n-1)},$$

whence

$$\operatorname*{Res}_{z=-n} \Gamma(z) = \frac{\Gamma(1)}{(-n)(-n+1)\cdots(-1)} = \frac{(-1)^n}{n!} \quad (n = 0, 1, 2, \dots). \tag{6.8}$$

Also, by repeated application of (6.6) we get $\Gamma(n+1) = n(n-1)\cdots 2 \cdot 1 \cdot \Gamma(1)$, i.e.,

$$\Gamma(n+1) = n! \qquad (n = 0, 1, 2, \dots). \tag{6.9}$$

The Euler beta-function is a function of two complex variables, defined by

$$B(x, y) = \int_0^1 t^{x-1}(1-t)^{y-1}\,\mathrm{d}t \qquad (x, y \in \mathbb{C},\ \operatorname{Re} x > 0,\ \operatorname{Re} y > 0) \tag{6.10}$$

and called by Legendre eulerian integral of the first kind.

The beta-function can easily be related to the gamma-function, as follows. With the substitution $t = u/(1+u)$ we get

$$B(x, y) = \int_0^\infty \frac{u^{x-1}}{(1+u)^{x+y}}\,\mathrm{d}u. \tag{6.11}$$

For any fixed $u \in (0, +\infty)$, with the substitution $(1+u)v = \vartheta$ we obtain

$$\int_0^\infty e^{-(1+u)v}\, v^{x+y-1}\,\mathrm{d}v = \int_0^\infty e^{-\vartheta}\left(\frac{\vartheta}{1+u}\right)^{x+y-1} \frac{\mathrm{d}\vartheta}{1+u}$$

$$= \frac{1}{(1+u)^{x+y}} \int_0^\infty e^{-\vartheta}\, \vartheta^{x+y-1}\,\mathrm{d}\vartheta = \frac{\Gamma(x+y)}{(1+u)^{x+y}}.$$

Multiplying by u^{x-1} and integrating in $0 < u < +\infty$, by (6.11) we find

$$\int_0^\infty du \int_0^\infty e^{-(1+u)v}\, u^{x-1} v^{x+y-1}\, dv = \Gamma(x+y) \int_0^\infty \frac{u^{x-1}}{(1+u)^{x+y}}\, du \tag{6.12}$$
$$= \Gamma(x+y) B(x,y),$$

where the repeated integral on the left-hand side of (6.12) can be inverted by absolute convergence.

For any fixed $v \in (0, +\infty)$, with the substitution $uv = \tau$ we get

$$\int_0^\infty e^{-uv}\, u^{x-1}\, du = \int_0^\infty e^{-\tau} \left(\frac{\tau}{v}\right)^{x-1} \frac{d\tau}{v} = v^{-x}\Gamma(x).$$

Hence, multiplying by $e^{-v}\, v^{x+y-1}$ and integrating in $0 < v < +\infty$,

$$\int_0^\infty dv \int_0^\infty e^{-(1+u)v}\, u^{x-1} v^{x+y-1}\, du$$
$$= \Gamma(x) \int_0^\infty e^{-v}\, v^{x+y-1} v^{-x}\, dv = \Gamma(x)\Gamma(y).$$

Comparing with (6.12) we conclude that

$$B(x,y) = \frac{\Gamma(x)\Gamma(y)}{\Gamma(x+y)}, \tag{6.13}$$

which yields the analytic continuation of $B(x,y)$ outside $\operatorname{Re} x > 0$ and $\operatorname{Re} y > 0$.

6.2 Euler's Limit Formula and Consequences

In this section we prove further important representations of $\Gamma(z)$ and functional equations, which we base upon Euler's limit formula, stated in the following

Theorem 6.1 (Euler) *For any* $z \in \mathbb{C} \setminus \{0, -1, -2, \dots\}$,

$$\Gamma(z) = \lim_{n\to\infty} \frac{n^z\, n!}{z(z+1)(z+2)\cdots(z+n)}. \tag{6.14}$$

Proof We first assume $\operatorname{Re} z > 0$. From (6.1) we get, for positive integer n,

$$\lim_{n\to\infty} \int_0^n \left(1-\frac{t}{n}\right)^n t^{z-1}\,dt \tag{6.15}$$

$$= \lim_{n\to\infty} \left(\int_0^n e^{-t} t^{z-1}\,dt - \int_0^n \left(e^{-t} - \left(1-\frac{t}{n}\right)^n\right) t^{z-1}\,dt \right)$$

$$= \Gamma(z) - \lim_{n\to\infty} \int_0^n \left(e^{-t} - \left(1-\frac{t}{n}\right)^n\right) t^{z-1}\,dt.$$

By (5.21),

$$\left| \int_0^n \left(e^{-t} - \left(1-\frac{t}{n}\right)^n\right) t^{z-1}\,dt \right| \le \int_0^n \frac{t^2}{n} e^{-t} t^{\operatorname{Re} z-1}\,dt = \frac{1}{n}\int_0^n e^{-t} t^{\operatorname{Re} z+1}\,dt$$

$$\le \frac{1}{n}\int_0^\infty e^{-t} t^{\operatorname{Re} z+1}\,dt \to 0 \quad (n\to\infty).$$

Therefore, by (6.15),

$$\Gamma(z) = \lim_{n\to\infty} \int_0^n \left(1-\frac{t}{n}\right)^n t^{z-1}\,dt. \tag{6.16}$$

With the substitution $t/n = \tau$ we get

$$\int_0^n \left(1-\frac{t}{n}\right)^n t^{z-1}\,dt = n^z \int_0^1 (1-\tau)^n \tau^{z-1}\,d\tau. \tag{6.17}$$

By repeated integrations by parts, since $\operatorname{Re} z > 0$,

$$\int_0^1 (1-\tau)^n \tau^{z-1}\,d\tau = \frac{n}{z}\int_0^1 (1-\tau)^{n-1} \tau^{z}\,d\tau = \frac{n}{z}\,\frac{n-1}{z+1}\int_0^1 (1-\tau)^{n-2} \tau^{z+1}\,d\tau$$

$$= \cdots = \frac{n!}{z(z+1)\cdots(z+n-1)}\int_0^1 \tau^{z+n-1}\,d\tau = \frac{n!}{z(z+1)\cdots(z+n)}.$$

Thus, by (6.17),

$$\int_0^n \left(1 - \frac{t}{n}\right)^n t^{z-1}\, dt = \frac{n^z\, n!}{z(z+1)\cdots(z+n)},$$

and Euler's limit formula (6.14) follows from (6.16).

If $\operatorname{Re} z \leq 0$, let $k \in \mathbb{N}$ be such that $-(k+1) < \operatorname{Re} z \leq -k$ with $z \neq -k$. From (6.7) we get

$$\Gamma(z) = \frac{1}{z(z+1)\cdots(z+k)}\, \Gamma(z+k+1).$$

Since $\operatorname{Re}(z+k+1) > 0$, by the previous argument we have

$$\begin{aligned}
\Gamma(z) &= \frac{1}{z(z+1)\cdots(z+k)} \lim_{n\to\infty} \frac{n^{z+k+1}\, n!}{(z+k+1)(z+k+2)\cdots(z+k+n+1)} \\
&= \lim_{n\to\infty} \left(\frac{n^z\, n!}{z(z+1)\cdots(z+n)} \cdot \frac{n^{k+1}}{(z+n+1)\cdots(z+n+k+1)} \right) \\
&= \lim_{n\to\infty} \frac{n^z\, n!}{z(z+1)\cdots(z+n)} \lim_{n\to\infty} \frac{n^{k+1}}{(n+z+1)\cdots(n+z+k+1)} \\
&= \lim_{n\to\infty} \frac{n^z\, n!}{z(z+1)\cdots(z+n)}.
\end{aligned}$$

□

Corollary 6.1 (Euler) *For any* $z \in \mathbb{C} \setminus \{0, -1, -2, \dots\}$,

$$\Gamma(z) = \frac{1}{z} \prod_{n=1}^{\infty} \left(1 + \frac{1}{n}\right)^z \left(1 + \frac{z}{n}\right)^{-1}. \tag{6.18}$$

Proof Since

$$\prod_{n=1}^{N-1} \left(1 + \frac{1}{n}\right) = \prod_{n=1}^{N-1} \frac{n+1}{n} = \frac{2}{1}\,\frac{3}{2}\,\frac{4}{3} \cdots \frac{N}{N-1} = N,$$

from (6.14) we get

$$\begin{aligned}
\Gamma(z) &= \lim_{N\to\infty} \frac{N^z}{z(1+z)(1+z/2)\cdots(1+z/N)} \qquad (6.19)\\
&= \frac{1}{z} \lim_{N\to\infty} \left(\prod_{n=1}^{N-1} \left(1 + \frac{1}{n}\right)^z \prod_{n=1}^{N} \left(1 + \frac{z}{n}\right)^{-1} \right)
\end{aligned}$$

$$= \frac{1}{z} \lim_{N\to\infty} \prod_{n=1}^{N} \left(1+\frac{1}{n}\right)^z \left(1+\frac{z}{n}\right)^{-1}$$
$$= \frac{1}{z} \prod_{n=1}^{\infty} \left(1+\frac{1}{n}\right)^z \left(1+\frac{z}{n}\right)^{-1}.$$

□

From (6.6) and (6.18),

$$\begin{aligned} z\,\Gamma(z)\,\Gamma(1-z) &= \Gamma(1+z)\,\Gamma(1-z) \qquad (6.20)\\ &= \frac{1}{1-z^2}\prod_{n=1}^{\infty}\left(1+\frac{1}{n}\right)^{1+z}\left(1+\frac{1}{n}\right)^{1-z}\left(1+\frac{1+z}{n}\right)^{-1}\left(1+\frac{1-z}{n}\right)^{-1}\\ &= \frac{1}{1-z^2}\prod_{n=1}^{\infty}\left(1+\frac{1}{n}\right)^2\left(\left(1+\frac{1}{n}\right)^2-\frac{z^2}{n^2}\right)^{-1}\\ &= \frac{1}{1-z^2}\prod_{n=1}^{\infty}\left(1-\frac{z^2}{(n+1)^2}\right)^{-1}\\ &= \prod_{n=1}^{\infty}\left(1-\frac{z^2}{n^2}\right)^{-1}. \end{aligned}$$

By (4.1) and (6.20),

$$z\,\Gamma(z)\,\Gamma(1-z) = \frac{\pi z}{\sin(\pi z)},$$

whence the second functional equation for the gamma-function:

$$\Gamma(z)\,\Gamma(1-z) = \frac{\pi}{\sin(\pi z)}, \qquad (6.21)$$

also called Euler's reflection formula. In particular, for $z = 1/2$, (6.21) yields $\Gamma(1/2)^2 = \pi$. By (6.1), $\Gamma(z) > 0$ for $z > 0$. Therefore

$$\Gamma(1/2) = \sqrt{\pi}. \qquad (6.22)$$

From (6.6) and (6.22) we obtain, by induction on n,

$$\Gamma\left(n+\frac{1}{2}\right) = \frac{(2n-1)!!}{2^n}\sqrt{\pi} \quad (n = 1, 2, 3, \dots). \qquad (6.23)$$

We now prove the third functional equation for the gamma-function, i.e., the following

Theorem 6.2 (Gauss' multiplication formula)

$$\prod_{k=0}^{n-1} \Gamma\Big(z+\frac{k}{n}\Big) = (2\pi)^{\frac{n-1}{2}}\, n^{\frac{1}{2}-nz}\, \Gamma(nz) \qquad (n=1,2,3,\dots). \tag{6.24}$$

Proof We write (6.14) in the form

$$\begin{aligned}\Gamma(z) &= \lim_{m\to\infty} \frac{m^z\,(m-1)!}{z(z+1)\cdots(z+m-1)} \lim_{m\to\infty} \frac{m}{z+m}\\ &= \lim_{m\to\infty} \frac{m^z\,(m-1)!}{z(z+1)\cdots(z+m-1)}.\end{aligned}$$

It follows that

$$\begin{aligned}\frac{n^{nz-1}}{\Gamma(nz)} \prod_{k=0}^{n-1} \Gamma\Big(z+\frac{k}{n}\Big) &= \frac{n^{nz-1}}{\lim\limits_{m\to\infty} \dfrac{(nm)^{nz}\,(nm-1)!}{nz(nz+1)\cdots(nz+nm-1)}}\\ &\quad \times \prod_{k=0}^{n-1} \lim_{m\to\infty} \frac{m^{z+k/n}\,(m-1)!}{(z+k/n)(z+k/n+1)\cdots(z+k/n+m-1)}\\ &= \lim_{m\to\infty} \frac{n^{nz-1}\, m^{nz+(n-1)/2}\,(m-1)!^n\, n^{nm}}{(nm)^{nz}\,(nm-1)!}\\ &= \lim_{m\to\infty} \frac{m^{(n-1)/2}\,(m-1)!^n\, n^{nm-1}}{(nm-1)!}.\end{aligned}$$

Thus the function

$$\frac{n^{nz-1}}{\Gamma(nz)} \prod_{k=0}^{n-1} \Gamma\Big(z+\frac{k}{n}\Big)$$

is independent of z, and hence equals identically its value at $z = 1/n$. Therefore

$$\frac{n^{nz-1}}{\Gamma(nz)} \prod_{k=0}^{n-1} \Gamma\Big(z+\frac{k}{n}\Big) = \prod_{k=1}^{n-1} \Gamma(k/n). \tag{6.25}$$

By (6.21),

$$\begin{aligned}\Big(\prod_{k=1}^{n-1} \Gamma(k/n)\Big)^2 &= \prod_{k=1}^{n-1} \Gamma(k/n)\,\Gamma(1-k/n)\\ &= \prod_{k=1}^{n-1} \pi\Big(\sin\frac{k\pi}{n}\Big)^{-1} = \pi^{n-1}\Big(\prod_{k=1}^{n-1} \sin\frac{k\pi}{n}\Big)^{-1}.\end{aligned}$$

Hence, by (6.25),

$$\frac{n^{nz-1}}{\Gamma(nz)} \prod_{k=0}^{n-1} \Gamma\left(z+\frac{k}{n}\right) = \pi^{(n-1)/2} \left(\prod_{k=1}^{n-1} \sin\frac{k\pi}{n}\right)^{-1/2}. \tag{6.26}$$

Since $x^n - 1 = (x-1)(x^{n-1} + x^{n-2} + \cdots + 1)$, the roots of the polynomial $x^{n-1} + x^{n-2} + \cdots + 1$ are the nth roots of 1 different from 1. Therefore

$$x^{n-1} + x^{n-2} + \cdots + 1 = \prod_{k=1}^{n-1} \left(x - e^{2k\pi i/n}\right)$$

whence, for $x = 1$,

$$\begin{aligned} n &= \prod_{k=1}^{n-1} \left(1 - e^{2k\pi i/n}\right) = \prod_{k=1}^{n-1} \left(-e^{k\pi i/n}\left(e^{k\pi i/n} - e^{-k\pi i/n}\right)\right) \\ &= \prod_{k=1}^{n-1} \left(-2i\, e^{k\pi i/n} \sin\frac{k\pi}{n}\right) = 2^{n-1} (-i)^{n-1} e^{i\frac{\pi}{n}\frac{n(n-1)}{2}} \prod_{k=1}^{n-1} \sin\frac{k\pi}{n} \\ &= 2^{n-1} \prod_{k=1}^{n-1} \sin\frac{k\pi}{n}. \end{aligned}$$

Consequently

$$\prod_{k=1}^{n-1} \sin\frac{k\pi}{n} = \frac{n}{2^{n-1}}, \tag{6.27}$$

and (6.24) follows from (6.26) and (6.27). □

In the special case $n = 2$, (6.24) is the Legendre duplication formula:

$$\Gamma(z)\,\Gamma(z+1/2) = \sqrt{\pi}\, 2^{1-2z}\, \Gamma(2z), \tag{6.28}$$

which for $z = 1/2$ yields again (6.22).

Theorem 6.3 (Weierstrass) *The Weierstrass factorization formula*

$$\frac{1}{\Gamma(z+1)} = \frac{1}{z\,\Gamma(z)} = e^{\gamma z} \prod_{n=1}^{\infty} \left(1+\frac{z}{n}\right) e^{-z/n}, \tag{6.29}$$

where γ is Euler's constant (5.18), holds for all $z \in \mathbb{C}$.

Proof By (6.19) and (5.18),

$$
\begin{aligned}
\frac{1}{\Gamma(z)} &= \lim_{N\to\infty}\left(z(1+z)(1+z/2)\cdots(1+z/N)e^{-z\log N}\right)\\
&= z\lim_{N\to\infty}\Bigg((1+z)e^{-z}(1+z/2)e^{-z/2}\cdots(1+z/N)e^{-z/N}\\
&\qquad\qquad\times\exp\left(z\left(\sum_{n=1}^{N}\frac{1}{n}-\log N\right)\right)\Bigg)\\
&= z\exp\left(z\lim_{N\to\infty}\left(\sum_{n=1}^{N}\frac{1}{n}-\log N\right)\right)\lim_{N\to\infty}\prod_{n=1}^{N}\left(1+\frac{z}{n}\right)e^{-z/n}\\
&= z\,e^{\gamma z}\prod_{n=1}^{\infty}\left(1+\frac{z}{n}\right)e^{-z/n},
\end{aligned}
$$

and (6.29) follows from (6.6). □

The sequence $z_n = -n$ $(n = 1, 2, 3, \ldots)$ has exponent of convergence $\beta = 1$, and the least integer satisfying (3.9) for $|z_n| = n$ is $p = 1$. Hence the infinite product

$$\prod_{n=1}^{\infty}\left(1+\frac{z}{n}\right)e^{-z/n}$$

in (6.29) is indeed the Weierstrass canonical product (3.11) for the sequence $z_n = -n$ and, by Theorem 3.5, is an entire function of order 1. Thus, by (6.29), $1/\Gamma(z+1)$ is an entire function with zeros $z = -1, -2, -3, \ldots$, corresponding to the poles of $\Gamma(z+1) = z\Gamma(z)$. With the notation of Corollary 3.2, for the entire function (6.29) we have $G(z) = \gamma z$, whence $\alpha = \beta = p = q = 1$.

Since $(z\Gamma(z))^{-1}$ is entire, we get

$$\Gamma(z) \neq 0 \quad \text{for all } z \in \mathbb{C}, \tag{6.30}$$

which is also an easy consequence of (6.21).

6.3 Stirling's Formula

Let $\mathbb{R}^- = (-\infty,\ 0\,]$ denote the set of real numbers ≤ 0, and let

$$\log\Gamma(z) = \log|\Gamma(z)| + i\arg\Gamma(z) \tag{6.31}$$

be the value of the logarithm of $\Gamma(z)$ with $\arg\Gamma(z) = 0$ for $z > 0$, and extended by continuity for $z \in \mathbb{C}\setminus\mathbb{R}^-$. By (6.30), and since the poles of $\Gamma(z)$ are in $\mathbb{R}^-$, in the simply connected open set $\mathbb{C} \setminus \mathbb{R}^-$ the function (6.31) has no branch points and therefore is one-valued and regular. We prove the following

Theorem 6.4 (Stirling's formula) *For any integer $r \geq 0$ and for any $\varepsilon > 0$,*

$$\log \Gamma(z) = \left(z - \frac{1}{2}\right) \log z - z + \log \sqrt{2\pi} + \sum_{h=1}^{r} \frac{B_{2h}}{(2h-1)2h} \frac{1}{z^{2h-1}} + O_{r,\varepsilon}\left(\frac{1}{|z|^{2r+1}}\right) \tag{6.32}$$

as $z \to \infty$ in the sector $-\pi + \varepsilon \leq \arg z \leq \pi - \varepsilon$. Here B_{2h} is the $(2h)$th Bernoulli number, $\log \Gamma(z)$ is as in (6.31), i.e. continuous in $\mathbb{C} \setminus \mathbb{R}^-$ with $\arg \Gamma(z) = 0$ for $z > 0$, and similarly $\log z$ is the principal logarithm, i.e., $\log z = \log |z| + i \arg z$ with $-\pi < \arg z < \pi$.

In particular we get the following improvement upon (5.6):

$$\log n! = \left(n + \frac{1}{2}\right) \log n - n + \log \sqrt{2\pi} + \sum_{h=1}^{r} \frac{B_{2h}}{(2h-1)2h} \frac{1}{n^{2h-1}} + O_r\left(\frac{1}{n^{2r+1}}\right) \quad (n \to \infty). \tag{6.33}$$

Proof By (6.6) and (6.9), $n! = \Gamma(n+1) = n\,\Gamma(n)$ whence $\log n! = \log \Gamma(n) + \log n$, so that (6.33) follows from (6.32) with $z = n \to \infty$.

Fix $z \in \mathbb{C} \setminus \mathbb{R}^-$, and let N be a positive integer. We have

$$\log \frac{N^z N!}{z(z+1)\cdots(z+N)} = z \log N - \log(z+N) - \sum_{k=1}^{N} \log\left(1 + \frac{z-1}{k}\right).$$

We apply to the last sum the Euler–MacLaurin summation formula (5.26) with $M = 1$, $f(t) = \log(1 + (z-1)/t) = \log(t + z - 1) - \log t$, and $q = 2r + 1$. A straightforward computation yields

$$\begin{aligned}
\log \frac{N^z N!}{z(z+1)\cdots(z+N)} &= \left(z - \frac{1}{2}\right) \log z - N \log\left(1 + \frac{z-1}{N}\right) \\
&\quad - \left(z - \frac{1}{2}\right) \log\left(1 + \frac{z-1}{N}\right) + \log\left(1 - \frac{z}{z+N}\right) \\
- \sum_{h=1}^{r} \frac{B_{2h}}{(2h-1)2h} &\left(\frac{1}{(z+N-1)^{2h-1}} - \frac{1}{N^{2h-1}}\right) + \sum_{h=1}^{r} \frac{B_{2h}}{(2h-1)2h} \frac{1}{z^{2h-1}} \\
- \sum_{h=1}^{r} \frac{B_{2h}}{(2h-1)2h} &- \frac{1}{2r+1} \int_1^N \frac{B_{2r+1}(\{t\})}{(z+t-1)^{2r+1}}\, dt + \frac{1}{2r+1} \int_1^N \frac{B_{2r+1}(\{t\})}{t^{2r+1}}\, dt.
\end{aligned}$$

Making in this formula $N \to \infty$, and assuming $r \ge 1$, we get by (6.14)

$$\log \Gamma(z) = \left(z - \frac{1}{2}\right) \log z - z + C_r + \sum_{h=1}^{r} \frac{B_{2h}}{(2h-1)2h} \frac{1}{z^{2h-1}} - \frac{1}{2r+1} \int_1^\infty \frac{B_{2r+1}(\{t\})}{(z+t-1)^{2r+1}} \, dt, \tag{6.34}$$

where

$$C_r = 1 - \sum_{h=1}^{r} \frac{B_{2h}}{(2h-1)2h} + \frac{1}{2r+1} \int_1^\infty \frac{B_{2r+1}(\{t\})}{t^{2r+1}} \, dt$$

is independent of z, and the integrals are absolutely convergent since $r \ge 1$ and $B_{2r+1}(\{t\})$ is bounded. Substituting $t - 1 = \tau$ and integrating by parts we obtain, using (4.25),

$$\begin{aligned} -\frac{1}{2r+1} \int_1^\infty \frac{B_{2r+1}(\{t\})}{(z+t-1)^{2r+1}} \, dt &= -\frac{1}{2r+1} \int_0^\infty \frac{B_{2r+1}(\{\tau\})}{(z+\tau)^{2r+1}} \, d\tau \\ &= \frac{B_{2r+2}}{(2r+1)(2r+2)} \frac{1}{z^{2r+1}} - \frac{1}{2r+2} \int_0^\infty \frac{B_{2r+2}(\{\tau\})}{(z+\tau)^{2r+2}} \, d\tau. \end{aligned} \tag{6.35}$$

Let $-\pi + \varepsilon \le \arg z \le \pi - \varepsilon$, and let $\vartheta = \pi - |\arg z|$, whence $\varepsilon \le \vartheta \le \pi$. By applying Carnot's theorem to the triangle of vertices 0, τ and $z + \tau$ we see that

$$|z+\tau|^2 = |z|^2 + \tau^2 - 2|z|\tau \cos \vartheta \ge |z|^2 + \tau^2 - 2|z|\tau \cos \varepsilon.$$

Therefore

$$\int_0^\infty \frac{B_{2r+2}(\{\tau\})}{(z+\tau)^{2r+2}} \, d\tau \ll \int_0^\infty \frac{d\tau}{|z+\tau|^{2r+2}} \le \int_0^\infty \frac{d\tau}{(|z|^2 + \tau^2 - 2|z|\tau \cos \varepsilon)^{r+1}}.$$

With the substitution $\tau = |z|(\cos \varepsilon + u \sin \varepsilon)$ the last integral becomes

$$\frac{1}{(|z| \sin \varepsilon)^{2r+1}} \int_{-\cot \varepsilon}^{+\infty} \frac{du}{(1+u^2)^{r+1}} < \frac{1}{(|z| \sin \varepsilon)^{2r+1}} \int_{-\infty}^{\infty} \frac{du}{(1+u^2)^{r+1}}.$$

Hence, by (6.35),

$$-\frac{1}{2r+1} \int_1^\infty \frac{B_{2r+1}(\{t\})}{(z+t-1)^{2r+1}} \, dt \ll_{r,\varepsilon} \frac{1}{|z|^{2r+1}} \quad \text{for } |\arg z| \le \pi - \varepsilon,$$

and (6.34) yields, for $r \geq 1$ and $|\arg z| \leq \pi - \varepsilon$,

$$\log \Gamma(z) = \left(z - \frac{1}{2}\right) \log z - z + C_r + \sum_{h=1}^{r} \frac{B_{2h}}{(2h-1)2h} \frac{1}{z^{2h-1}} + O_{r,\varepsilon}\left(\frac{1}{|z|^{2r+1}}\right). \tag{6.36}$$

Since

$$\frac{1}{|z|^{2h-1}} \ll \frac{1}{|z|} \quad \text{for } h = 1, \ldots, r, r+1 \text{ and } |z| \to \infty,$$

by (6.36) we get

$$\log \Gamma(z) = \left(z - \frac{1}{2}\right) \log z - z + C_r + O_{r,\varepsilon}\left(\frac{1}{|z|}\right). \tag{6.37}$$

In particular, for $z = n \to \infty$,

$$\log n! = \log \Gamma(n) + \log n = \left(n + \frac{1}{2}\right) \log n - n + C_r + O_r\left(\frac{1}{n}\right).$$

Comparing this asymptotic formula with (5.6) we get, for all $r \geq 1$,

$$C_r = \log \sqrt{2\pi}, \tag{6.38}$$

and (6.32) follows from (6.36) and (6.38). Moreover, by (6.37) with a fixed $r \geq 1$ and by (6.38),

$$\log \Gamma(z) = \left(z - \frac{1}{2}\right) \log z - z + \log \sqrt{2\pi} + O_\varepsilon\left(\frac{1}{|z|}\right), \tag{6.39}$$

which shows that (6.32) holds also for $r = 0$. □

We incidentally remark that (6.38) can also be obtained by taking the logarithm of the Legendre duplication formula (6.28), and then replacing $\log \Gamma(z)$, $\log \Gamma(z + 1/2)$ and $\log \Gamma(2z)$ with the values given by (6.37).

With the same meaning as in (5.31), Stirling's formula (6.32) can symbolically be written as

$$\log \Gamma(z) \sim \left(z - \frac{1}{2}\right) \log z - z + \log \sqrt{2\pi} + \sum_{h=1}^{\infty} \frac{B_{2h}}{(2h-1)2h} \frac{1}{z^{2h-1}},$$

where, by (5.13), the series diverges for every z.

Stirling's formula (6.32) can be further generalized as follows.

Corollary 6.2 *For any integer $s \geq 0$, for any $\eta > 0$ and for any arbitrarily small $\varepsilon > 0$,*

$$\begin{aligned}\log \Gamma(z+\alpha) = \; & \big(z + B_1(\alpha)\big)\log z - z + \log\sqrt{2\pi} \\ & + \sum_{k=1}^{s} (-1)^{k+1} \frac{B_{k+1}(\alpha)}{k(k+1)} \frac{1}{z^k} + O_{s,\eta,\varepsilon}\Big(\frac{1}{|z|^{s+1}}\Big)\end{aligned} \tag{6.40}$$

uniformly for $z, \alpha \in \mathbb{C}$ satisfying $|\alpha| \leq \eta$, $|z| \geq |\alpha| + \varepsilon$, $|\arg(z+\alpha)| \leq \pi - \varepsilon$, where $B_1(\alpha)$, $B_2(\alpha)$, $B_3(\alpha)$, ... are the Bernoulli polynomials.

Proof Owing to (4.9), we can write (6.32) in the form

$$\begin{aligned}\log \Gamma(z) = \; & \Big(z - \frac{1}{2}\Big)\log z - z + \log\sqrt{2\pi} \\ & + \sum_{k=1}^{s} (-1)^{k+1} \frac{B_{k+1}}{k(k+1)} \frac{1}{z^k} + O_{s,\varepsilon}\Big(\frac{1}{|z|^{s+1}}\Big)\end{aligned} \tag{6.41}$$

for any integer $s \geq 0$, as $z \to \infty$ in the sector $|\arg z| \leq \pi - \varepsilon$.

In (6.41) we change z to $z + \alpha$. We get

$$\begin{aligned}\log \Gamma(z+\alpha) = \; & \Big(z + \alpha - \frac{1}{2}\Big)\Big(\log z + \log\Big(1 + \frac{\alpha}{z}\Big)\Big) - z - \alpha + \log\sqrt{2\pi} \\ & + \sum_{k=1}^{s} (-1)^{k+1} \frac{B_{k+1}}{k(k+1)} z^{-k}\Big(1 + \frac{\alpha}{z}\Big)^{-k} + O\Big(\frac{1}{|z|^{s+1}}\Big).\end{aligned}$$

Since $|\alpha| \leq \eta$ and $|\alpha/z| \leq 1 - \varepsilon/(\eta + \varepsilon)$, the Taylor expansions of $\log(1 + \alpha/z)$ and $(1 + \alpha/z)^{-k}$ yield

$$\begin{aligned}\log \Gamma(z+\alpha) = \; & \Big(z + \alpha - \frac{1}{2}\Big)\log z - z + \log\sqrt{2\pi} - \alpha + \sum_{m=0}^{s} (-1)^m \frac{\alpha^{m+1}}{(m+1)z^m} \\ & + \Big(\alpha - \frac{1}{2}\Big)\sum_{m=1}^{s} (-1)^{m+1} \frac{\alpha^m}{m z^m} + \sum_{k=1}^{s} (-1)^{k+1} \frac{B_{k+1}}{k(k+1)} \sum_{m=k}^{s} \binom{-k}{m-k} \frac{\alpha^{m-k}}{z^m} \\ & + O\Big(\frac{1}{|z|^{s+1}}\Big) \\ = \; & \Big(z + \alpha - \frac{1}{2}\Big)\log z - z + \log\sqrt{2\pi} + \sum_{m=1}^{s} (-1)^m \frac{\alpha^{m+1}}{(m+1)z^m} \\ & + \sum_{m=1}^{s} (-1)^{m+1} \frac{\alpha^{m+1} - \alpha^m/2}{m z^m} + \sum_{m=1}^{s} \frac{1}{z^m} \sum_{k=1}^{m} (-1)^{k+1} \binom{-k}{m-k} \frac{B_{k+1}\alpha^{m-k}}{k(k+1)} \\ & + O\Big(\frac{1}{|z|^{s+1}}\Big)\end{aligned}$$

$$= \left(z+\alpha-\frac{1}{2}\right)\log z - z + \log\sqrt{2\pi} + \sum_{m=1}^{s}(-1)^{m+1}\left(\frac{\alpha^{m+1}-(m+1)\alpha^m/2}{m(m+1)}\right.$$
$$\left. + \sum_{k=1}^{m}(-1)^{m-k}\binom{-k}{m-k}\frac{B_{k+1}\alpha^{m-k}}{k(k+1)}\right)\frac{1}{z^m} + O\left(\frac{1}{|z|^{s+1}}\right).$$

Since

$$(-1)^{m-k}\binom{-k}{m-k} = \binom{m-1}{m-k} = \binom{m-1}{k-1},$$

we get

$$\frac{(-1)^{m-k}}{k(k+1)}\binom{-k}{m-k} = \frac{1}{m(m+1)}\frac{m(m+1)}{k(k+1)}\binom{m-1}{k-1} = \frac{1}{m(m+1)}\binom{m+1}{k+1}.$$

Therefore, using (4.20),

$$\log\Gamma(z+\alpha) = \left(z+\alpha-\frac{1}{2}\right)\log z - z + \log\sqrt{2\pi}$$
$$+\sum_{m=1}^{s}\frac{(-1)^{m+1}}{m(m+1)}\left(\alpha^{m+1}-\frac{m+1}{2}\alpha^m + \sum_{k=1}^{m}\binom{m+1}{k+1}B_{k+1}\alpha^{m-k}\right)\frac{1}{z^m}$$
$$+O\left(\frac{1}{|z|^{s+1}}\right)$$
$$= (z+B_1(\alpha))\log z - z + \log\sqrt{2\pi}$$
$$+\sum_{m=1}^{s}(-1)^{m+1}\frac{B_{m+1}(\alpha)}{m(m+1)}\frac{1}{z^m} + O\left(\frac{1}{|z|^{s+1}}\right).$$

□

Note that in the special case $\alpha = 0$ (6.40) becomes (6.41), i.e., (6.32).

Corollary 6.3 *Let $x_1, x_2 \in \mathbb{R}$ be fixed, $x_1 \le x_2$. For $x+iy$ in the vertical strip $x_1 \le x \le x_2$, the following asymptotic formula holds:*

$$|\Gamma(x+iy)| = \sqrt{2\pi}\,|y|^{x-\frac{1}{2}}\,e^{-\frac{\pi}{2}|y|}\left(1+O_{x_1,x_2}\left(\frac{1}{|y|}\right)\right) \quad (y\to\pm\infty). \tag{6.42}$$

Proof In (6.40) we take $s=0$, $\eta=\max\{|x_1|,|x_2|\}$, $\alpha=x$, $z=iy$. For sufficiently large $|y|$ we get

$$\log\Gamma(x+iy) = \left(x+iy-\frac{1}{2}\right)\log(iy) - iy + \log\sqrt{2\pi} + O\left(\frac{1}{|y|}\right)$$
$$= \left(x-\frac{1}{2}+iy\right)\left(\log|y| + i\frac{\pi}{2}\operatorname{sgn} y\right) - iy + \log\sqrt{2\pi} + O\left(\frac{1}{|y|}\right),$$

whence

$$\begin{aligned}\log|\Gamma(x+iy)| &= \operatorname{Re}\left(\log\Gamma(x+iy)\right)\\ &= \left(x-\frac{1}{2}\right)\log|y| - \frac{\pi}{2}|y| + \log\sqrt{2\pi} + O\left(\frac{1}{|y|}\right).\end{aligned}$$

Taking exponentials we get (6.42). □

6.4 The Psi-Function

The logarithmic derivative of the gamma-function is traditionally denoted by ψ:

$$\psi(z) := \frac{\Gamma'(z)}{\Gamma(z)}, \tag{6.43}$$

and is called by some authors the digamma-function. Some formulae for $\Gamma(z)$ turn into simpler formulae for $\psi(z)$.

By taking logarithms and then differentiating, the functional equations (6.6), (6.21) and (6.24) for the gamma-function yield, respectively,

$$\psi(1+z) = \psi(z) + \frac{1}{z}, \tag{6.44}$$

$$\psi(1-z) = \psi(z) + \pi\cot(\pi z) \tag{6.45}$$

and

$$\psi(nz) = \sum_{k=0}^{n-1}\frac{1}{n}\psi\left(z+\frac{k}{n}\right) + \log n \quad (n = 1, 2, 3, \dots). \tag{6.46}$$

By (6.30), and since the poles $z = 0, -1, -2, \dots$ of $\Gamma(z)$ are all simple, $\psi(z)$ is regular in $\mathbb{C}\setminus\{0, -1, -2, \dots\}$, with simple poles at $z = 0, -1, -2, \dots$, all with residue -1. Taking the logarithm of (6.29) we get

$$\log\Gamma(z) = -\log z - \gamma z + \sum_{n=1}^{\infty}\left(\frac{z}{n} - \log\left(1+\frac{z}{n}\right)\right), \tag{6.47}$$

where γ is Euler's constant. Differentiating (6.47) we obtain

$$\psi(z) = -\gamma - \frac{1}{z} + \sum_{n=1}^{\infty}\left(\frac{1}{n} - \frac{1}{n+z}\right) = -\gamma - \frac{1}{z} + \sum_{n=1}^{\infty}\frac{z}{n(n+z)}, \tag{6.48}$$

where term-by-term differentiation is justified by Theorems 2.1 and 2.2.

By (6.44) and (6.48),

$$\psi(1+z) = -\gamma + \sum_{n=1}^{\infty}\frac{z}{n(n+z)}. \tag{6.49}$$

For any $a, b \in \mathbb{C} \setminus \{0, -1, -2, \dots\}$ we have, by (6.48),

$$\psi(b) - \psi(a) = \frac{1}{a} - \frac{1}{b} + \sum_{n=1}^{\infty}\left(\frac{1}{n+a} - \frac{1}{n+b}\right) = \sum_{n=0}^{\infty}\left(\frac{1}{n+a} - \frac{1}{n+b}\right), \tag{6.50}$$

whence, for $a \neq b$,

$$\frac{\psi(b) - \psi(a)}{b-a} = \sum_{n=0}^{\infty}\frac{1}{(n+a)(n+b)}.$$

Thus, using the function ψ, one can sum a series of the type

$$\sum_{n=0}^{\infty}\frac{1}{P(n)} \tag{6.51}$$

where P is a polynomial of degree 2 with distinct roots $\notin \mathbb{N}$. Since, by (6.48),

$$\psi'(z) = \frac{1}{z^2} + \sum_{n=1}^{\infty}\frac{1}{(n+z)^2} = \sum_{n=0}^{\infty}\frac{1}{(n+z)^2}, \tag{6.52}$$

for $a \neq 0, -1, -2, \dots$ we get

$$\psi'(a) = \sum_{n=0}^{\infty}\frac{1}{(n+a)^2},$$

which yields the sum of (6.51) if P has a double root.

Plainly this argument can be iterated to sum (6.51) if P is a polynomial of degree > 2. For instance, if $\deg P = 3$ and the roots of P are distinct and $\notin \mathbb{N}$, we get

$$\begin{aligned}\sum_{n=0}^{\infty}\frac{1}{(n+a)(n+b)(n+c)} &= \frac{1}{b-a}\sum_{n=0}^{\infty}\Big(\frac{1}{n+a}-\frac{1}{n+b}\Big)\frac{1}{n+c}\\ &= \frac{1}{b-a}\Big(\sum_{n=0}^{\infty}\frac{1}{(n+a)(n+c)}-\sum_{n=0}^{\infty}\frac{1}{(n+b)(n+c)}\Big)\\ &= \frac{1}{b-a}\Big(\frac{\psi(c)-\psi(a)}{c-a}-\frac{\psi(c)-\psi(b)}{c-b}\Big)\\ &= \frac{\psi(a)}{(b-a)(a-c)}+\frac{\psi(b)}{(c-b)(b-a)}+\frac{\psi(c)}{(a-c)(c-b)}.\end{aligned}$$

Since $\Gamma(1)=1$, from (6.43) and (6.48) we get

$$\psi(1) = \Gamma'(1) = -\gamma - 1 + \sum_{n=1}^{\infty}\Big(\frac{1}{n}-\frac{1}{n+1}\Big),$$

and clearly

$$\sum_{n=1}^{\infty}\Big(\frac{1}{n}-\frac{1}{n+1}\Big) = 1.$$

Therefore

$$\psi(1) = \Gamma'(1) = -\gamma. \tag{6.53}$$

From (6.44) and (6.53) we obtain, by induction on n,

$$\psi(n) = -\gamma + \sum_{k=1}^{n-1}\frac{1}{k} \qquad (n = 1, 2, 3, \dots). \tag{6.54}$$

We now apply (3.28) to the entire function (6.29). For $F(z) = 1/\Gamma(z+1)$ we get

$$\frac{F'(z)}{F(z)} = -\frac{\Gamma'(z+1)}{\Gamma(z+1)} = -\psi(z+1).$$

Hence, by (3.28),

$$\sum_{n=1}^{\infty}(-n)^{-k} = (-1)^k\zeta(k) = \frac{1}{(k-1)!}\psi^{(k-1)}(1) \qquad (k = 2, 3, 4, \dots) \tag{6.55}$$

where ζ is the Riemann zeta-function. By (6.53) and (6.55) we get the Taylor expansion

$$\psi(1+z) = -\gamma + \sum_{k=2}^{\infty}(-1)^k\zeta(k)z^{k-1} \qquad (|z| < 1). \tag{6.56}$$

We remark that the Taylor series (6.56) easily yields another proof of Euler's formulae (4.16) for $\zeta(2k)$. From (6.44) and (6.45) we have

$$\psi(1+z) - \psi(1-z) = \frac{1}{z} - \pi \cot(\pi z),$$

whence, by (4.15),

$$\psi(1+z) - \psi(1-z) = \sum_{k=1}^{\infty} (-1)^{k-1} (2\pi)^{2k} B_{2k} \frac{z^{2k-1}}{(2k)!} \qquad (|z| < 1). \tag{6.57}$$

Changing z to $-z$ in (6.56),

$$\psi(1-z) = -\gamma - \sum_{k=2}^{\infty} \zeta(k) z^{k-1}.$$

Subtracting this formula from (6.56) we get

$$\psi(1+z) - \psi(1-z) = \sum_{k=2}^{\infty} \big(1 + (-1)^k\big) \zeta(k) z^{k-1} = 2 \sum_{k=1}^{\infty} \zeta(2k) z^{2k-1}. \tag{6.58}$$

Comparing (6.57) and (6.58) we obtain $2\,\zeta(2k) = (-1)^{k-1}(2\pi)^{2k} B_{2k}/(2k)!$, i.e., (4.16).

Lemma 6.1 *For fixed $q \in \mathbb{R}$ and $\varepsilon > 0$, let $f(z)$ be regular and satisfy*

$$f(z) \ll_{q,\varepsilon} \frac{1}{|z|^q} \qquad (z \to \infty) \tag{6.59}$$

in the sector $-\pi + \varepsilon/2 \le \arg z \le \pi - \varepsilon/2$. Then

$$f'(z) \ll_{q,\varepsilon} \frac{1}{|z|^{q+1}} \qquad (z \to \infty)$$

for $-\pi + \varepsilon \le \arg z \le \pi - \varepsilon$.

Proof Let $z \in \mathbb{C}$ satisfy $-\pi + \varepsilon \le \arg z \le \pi - \varepsilon$, and let t be on the circumference $\gamma_{z,\varepsilon}$ of centre z and radius $|z| \sin(\varepsilon/2)$. A picture shows that $|\arg t\,| \le \pi - \varepsilon/2$ and

$$|t| \ge |z| - |z| \sin\frac{\varepsilon}{2} = |z|\Big(1 - \sin\frac{\varepsilon}{2}\Big). \tag{6.60}$$

From Cauchy's integral formula we get

$$f'(z) = \frac{1}{2\pi i} \oint_{\gamma_{z,\varepsilon}} \frac{f(t)}{(t-z)^2}\, \mathrm{d}t,$$

whence, by (6.59) and (6.60),

$$\begin{aligned}|f'(z)| &\le \frac{1}{2\pi}\oint_{\gamma_{z,\varepsilon}} \frac{|f(t)|}{(|z|\sin(\varepsilon/2))^2}\,|\mathrm{d}t| \ll_{q,\varepsilon} \frac{1}{2\pi(|z|\sin(\varepsilon/2))^2}\oint_{\gamma,\varepsilon}\frac{|\mathrm{d}t|}{|t|^q}\\ &\le \frac{1}{2\pi(|z|\sin(\varepsilon/2))^2\,|z|^q\,(1-\sin(\varepsilon/2))^q}\,2\pi|z|\sin\frac{\varepsilon}{2}\\ &= \frac{1}{|z|^{q+1}\sin(\varepsilon/2)\,(1-\sin(\varepsilon/2))^q} \ll_{q,\varepsilon} \frac{1}{|z|^{q+1}}.\end{aligned}$$

□

Theorem 6.5 (Stirling's formula for ψ) *For any integer $s \ge 0$, for any $\eta > 0$ and for any arbitrarily small $\varepsilon > 0$,*

$$\begin{aligned}\psi(z+\alpha) &= \frac{\Gamma'(z+\alpha)}{\Gamma(z+\alpha)} \qquad (6.61)\\ &= \log z + \frac{B_1(\alpha)}{z} + \sum_{k=1}^{s}(-1)^k\frac{B_{k+1}(\alpha)}{k+1}\frac{1}{z^{k+1}} + O_{s,\eta,\varepsilon}\left(\frac{1}{|z|^{s+2}}\right)\end{aligned}$$

uniformly for $z, \alpha \in \mathbb{C}$ satisfying $|\alpha| \le \eta$, $|z| \ge |\alpha| + \varepsilon$, $|\arg(z+\alpha)| \le \pi - \varepsilon$.

In the special case $\alpha = 0$ we get, for any integer $r \ge 0$,

$$\psi(z) = \log z - \frac{1}{2z} - \sum_{h=1}^{r}\frac{B_{2h}}{2h}\frac{1}{z^{2h}} + O_{r,\varepsilon}\left(\frac{1}{|z|^{2r+2}}\right) \qquad (6.62)$$

as $z \to \infty$ in the sector $-\pi + \varepsilon \le \arg z \le \pi - \varepsilon$.

Proof Let

$$f(z) := \log\Gamma(z) - \left(z - \frac{1}{2}\right)\log z + z - \log\sqrt{2\pi} - \sum_{h=1}^{r}\frac{B_{2h}}{(2h-1)2h}\frac{1}{z^{2h-1}}$$

for $|\arg z| \le \pi - \varepsilon/2$. By (6.32) we have

$$f(z) \ll_{r,\varepsilon} \frac{1}{|z|^{2r+1}} \qquad (z \to \infty).$$

Since

$$f'(z) = \psi(z) - \log z + \frac{1}{2z} + \sum_{h=1}^{r}\frac{B_{2h}}{2h}\frac{1}{z^{2h}},$$

(6.62) follows from Lemma 6.1.

Similarly, applying Lemma 6.1 (with z replaced by $z+\alpha$) to the function

$$\log \Gamma(z+\alpha) - \big(z + B_1(\alpha)\big) \log z + z - \log\sqrt{2\pi} - \sum_{k=1}^{s} (-1)^{k+1} \frac{B_{k+1}(\alpha)}{k(k+1)} \frac{1}{z^k},$$

we get (6.61) from (6.40). □

Theorem 6.6 (Gauss) *For any* $z \in \mathbb{C}$ *such that* $\operatorname{Re} z > 0$,

$$\psi(z) = \int_0^{+\infty} \left(\frac{e^{-t}}{t} - \frac{e^{-tz}}{1-e^{-t}} \right) \mathrm{d}t. \tag{6.63}$$

Proof For z in the half-plane $\operatorname{Re} z > 0$ we have

$$\begin{aligned} \int_0^{+\infty} e^{-tz}\,\mathrm{d}t &= \lim_{\tau\to+\infty} \int_0^{\tau} e^{-tz}\,\mathrm{d}t = \frac{1}{z} \lim_{\tau\to+\infty} \int_0^{\tau z} e^{-u}\,\mathrm{d}u \\ &= \frac{1}{z} \lim_{\tau\to+\infty} \left(1 - e^{-\tau z}\right) = \frac{1}{z}, \end{aligned}$$

because

$$\left| e^{-\tau z} \right| = e^{-\tau \operatorname{Re} z} \to 0 \qquad (\tau \to +\infty). \tag{6.64}$$

Changing z to $z+m$ we get

$$\int_0^{+\infty} e^{-t(z+m)}\,\mathrm{d}t = \frac{1}{z+m} \qquad (m = 0, 1, 2, \ldots). \tag{6.65}$$

Hence, by (6.48),

$$\begin{aligned} \psi(z) &= -\gamma - \int_0^{\infty} e^{-tz}\,\mathrm{d}t + \lim_{n\to\infty} \sum_{m=1}^{n} \left(\int_0^{\infty} e^{-tm}\,\mathrm{d}t - \int_0^{\infty} e^{-t(z+m)}\,\mathrm{d}t \right) \\ &= -\gamma + \lim_{n\to\infty} \int_0^{\infty} \left(\sum_{m=1}^{n} e^{-tm} - \sum_{m=1}^{n} e^{-t(z+m)} - e^{-tz} \right) \mathrm{d}t \\ &= -\gamma + \lim_{n\to\infty} \int_0^{\infty} \frac{e^{-t} - e^{-tz} - e^{-t(n+1)} + e^{-t(z+n+1)}}{1-e^{-t}}\,\mathrm{d}t. \end{aligned}$$

Combining this formula with (5.24) we obtain

$$\psi(z) = \int_0^\infty \left(\frac{e^{-t}}{t} - \frac{e^{-tz}}{1-e^{-t}}\right) dt - \lim_{n\to\infty} \int_0^\infty \frac{1-e^{-tz}}{1-e^{-t}} e^{-t(n+1)}\, dt. \tag{6.66}$$

Since

$$\lim_{t\to 0} \frac{1-e^{-tz}}{1-e^{-t}} = z$$

and, by (6.64),

$$\lim_{t\to +\infty} \frac{1-e^{-tz}}{1-e^{-t}} = 1,$$

the function $|(1-e^{-tz})/(1-e^{-t})|$ is bounded for $t \in (0, +\infty)$. Thus there exists a constant $K = K(z) > 0$, independent of t, such that

$$\left|\frac{1-e^{-tz}}{1-e^{-t}}\right| < K \quad \text{for all } t > 0.$$

Therefore, by (6.65),

$$\left|\int_0^\infty \frac{1-e^{-tz}}{1-e^{-t}} e^{-t(n+1)}\, dt\right| \le K \int_0^\infty e^{-t(n+1)}\, dt = \frac{K}{n+1} \to 0 \quad (n \to \infty), \tag{6.67}$$

and (6.63) follows from (6.66) and (6.67). □

6.5 Binet's Integral Formulae

It is natural to ask whether, for suitable fixed z, in Stirling's formula (6.39) the remainder term

$$\log \Gamma(z) - \left(z - \frac{1}{2}\right) \log z + z - \log \sqrt{2\pi}$$

can be given a closed integral form independent of Bernoulli numbers and polynomials. Binet gave an affirmative answer to this question via the function $\psi(z)$, with two different integral formulae valid for any $z \in \mathbb{C}$ in the half-plane $\operatorname{Re} z > 0$ (so that in Binet's formulae the ε appearing in (6.39) is immaterial).

Theorem 6.7 (Binet's first integral formula) *For any* $z \in \mathbb{C}$ *such that* $\operatorname{Re} z > 0$,

$$\log \Gamma(z) = \left(z - \frac{1}{2}\right) \log z - z + \log \sqrt{2\pi} + \int_0^{+\infty} \left(\frac{1}{2} - \frac{1}{t} + \frac{1}{e^t - 1}\right) \frac{e^{-tz}}{t}\, dt.$$

Proof Integrating along the segment of endpoints 1 and z we get by (6.65)

$$\log z = \int_1^z \frac{\mathrm{d}u}{u} = \int_1^z \mathrm{d}u \int_0^{+\infty} e^{-tu}\,\mathrm{d}t = \int_0^{+\infty} \mathrm{d}t \int_1^z e^{-tu}\,\mathrm{d}u = \int_0^{+\infty} \frac{e^{-t} - e^{-tz}}{t}\,\mathrm{d}t, \tag{6.68}$$

where the interchange of integrations is justified by

$$\int_1^z |\mathrm{d}u| \int_0^{+\infty} |e^{-tu}|\,\mathrm{d}t = \int_1^z |\mathrm{d}u| \int_0^{+\infty} e^{-t\,\mathrm{Re}\,u}\,\mathrm{d}t = \int_1^z \frac{|\mathrm{d}u|}{\mathrm{Re}\,u} \le \frac{|z-1|}{\min\{1, \mathrm{Re}\,z\}} < +\infty.$$

Thus, by (6.63), (6.65) and (6.68),

$$\begin{aligned}\psi(z) = \frac{\Gamma'(z)}{\Gamma(z)} &= \int_0^\infty \left(\frac{e^{-t}}{t} - e^{-tz}\left(1 + \frac{1}{e^t - 1}\right)\right)\mathrm{d}t = \int_0^\infty \left(\frac{e^{-t}}{t} - e^{-tz} - \frac{e^{-tz}}{e^t - 1}\right)\mathrm{d}t \\ &= \int_0^\infty \frac{e^{-t} - e^{-tz}}{t}\,\mathrm{d}t - \frac{1}{2}\int_0^\infty e^{-tz}\,\mathrm{d}t - \int_0^\infty \left(\frac{1}{2} - \frac{1}{t} + \frac{1}{e^t - 1}\right)e^{-tz}\,\mathrm{d}t \\ &= \log z - \frac{1}{2z} - \int_0^\infty \left(\frac{1}{2} - \frac{1}{t} + \frac{1}{e^t - 1}\right)e^{-tz}\,\mathrm{d}t.\end{aligned}$$

We use this formula to integrate $\psi(u) = \Gamma'(u)/\Gamma(u)$ along the segment of endpoints 1 and z. We get

$$\begin{aligned}\log \Gamma(z) &= \int_1^z \log u\,\mathrm{d}u - \frac{1}{2}\int_1^z \frac{\mathrm{d}u}{u} - \int_1^z \mathrm{d}u \int_0^\infty \left(\frac{1}{2} - \frac{1}{t} + \frac{1}{e^t - 1}\right)e^{-tu}\,\mathrm{d}t \\ &= \left(z - \frac{1}{2}\right)\log z - z + 1 - \int_0^\infty \left(\frac{1}{2} - \frac{1}{t} + \frac{1}{e^t - 1}\right)\mathrm{d}t \int_1^z e^{-tu}\,\mathrm{d}u,\end{aligned}$$

where the interchange of integrations is allowed as in (6.68) because

$$\left|\frac{1}{2} - \frac{1}{t} + \frac{1}{e^t - 1}\right|,$$

by (4.8), tends to 0 as $t \to 0$ and to $1/2$ as $t \to +\infty$, and hence is bounded for $t \in (0, +\infty)$. Therefore

$$\log \Gamma(z) = \left(z - \frac{1}{2}\right) \log z - z + 1 - \int_0^\infty \left(\frac{1}{2} - \frac{1}{t} + \frac{1}{e^t - 1}\right) \frac{e^{-t} - e^{-tz}}{t} \, \mathrm{d}t \quad (6.69)$$

$$= \left(z - \frac{1}{2}\right) \log z - z + C + \int_0^\infty \left(\frac{1}{2} - \frac{1}{t} + \frac{1}{e^t - 1}\right) \frac{e^{-tz}}{t} \, \mathrm{d}t,$$

where

$$C = 1 - \int_0^\infty \left(\frac{1}{2} - \frac{1}{t} + \frac{1}{e^t - 1}\right) \frac{e^{-t}}{t} \, \mathrm{d}t \quad (6.70)$$

is a constant. Note that the integrals in (6.69) and (6.70) are absolutely convergent because, again by (4.8),

$$\lim_{t \to 0} \left| \left(\frac{1}{2} - \frac{1}{t} + \frac{1}{e^t - 1}\right) \frac{1}{t} \right| = \frac{1}{12}.$$

Moreover, denoting

$$M = \max_{t \geq 0} \left| \left(\frac{1}{2} - \frac{1}{t} + \frac{1}{e^t - 1}\right) \frac{1}{t} \right|,$$

we get, by (6.65),

$$\left| \int_0^\infty \left(\frac{1}{2} - \frac{1}{t} + \frac{1}{e^t - 1}\right) \frac{e^{-tz}}{t} \, \mathrm{d}t \right| \leq M \int_0^\infty e^{-t \operatorname{Re} z} \, \mathrm{d}t = \frac{M}{\operatorname{Re} z}.$$

Hence for $z > 0$, $z \to +\infty$, (6.69) yields

$$\log \Gamma(z) = \left(z - \frac{1}{2}\right) \log z - z + C + O\left(\frac{1}{z}\right).$$

Comparing this asymptotic formula with (6.39) we obtain $C = \log \sqrt{2\pi}$, and the theorem follows from (6.69). □

Theorem 6.8 (Binet's second integral formula) *For any* $z \in \mathbb{C}$ *such that* $\operatorname{Re} z > 0$,

$$\log \Gamma(z) = \left(z - \frac{1}{2}\right) \log z - z + \log \sqrt{2\pi} + 2 \int_0^{+\infty} \frac{\arctan(t/z)}{e^{2\pi t} - 1} \, \mathrm{d}t,$$

where the branch of arc tangent is defined by $\arctan(t/z) = \int_0^{t/z} \mathrm{d}u/(1 + u^2)$ *along a path from* 0 *to* t/z *in the half-plane* $\operatorname{Re} u > 0$.

Proof We apply Theorem 5.4 with the function $f(v) := (z+v)^{-2}$ and with $r = 0$. Then (5.34) yields, for any integer $s > 0$,

$$\frac{1}{2z^2} + \sum_{k=1}^{s-1} \frac{1}{(z+k)^2} + \frac{1}{2(z+s)^2} \tag{6.71}$$

$$= \frac{1}{z} - \frac{1}{z+s} + i\int_0^{+\infty} \left(\frac{1}{(z+it)^2} - \frac{1}{(z-it)^2} \right) \frac{dt}{e^{2\pi t}-1}$$

$$- i\int_0^{+\infty} \left(\frac{1}{(z+s+it)^2} - \frac{1}{(z+s-it)^2} \right) \frac{dt}{e^{2\pi t}-1}.$$

Since

$$\frac{1}{(z+it)^2} - \frac{1}{(z-it)^2} = -\frac{4itz}{(z^2+t^2)^2} \tag{6.72}$$

we have

$$\left| \frac{1}{(z+s+it)^2} - \frac{1}{(z+s-it)^2} \right| = \frac{4t|z+s|}{|(z+s)^2+t^2|^2}.$$

Plainly $\lim_{s\to+\infty} \arg(z+s) = 0$, so that for any $s > s_0(z)$ we get $|\arg(z+s)| < \pi/4$, whence $\mathrm{Re}\left((z+s)^2\right) > 0$, $|(z+s)^2+t^2| \geq |z+s|^2$. Thus for $s > s_0(z)$ we have

$$\left| \int_0^{+\infty} \left(\frac{1}{(z+s+it)^2} - \frac{1}{(z+s-it)^2} \right) \frac{dt}{e^{2\pi t}-1} \right| \tag{6.73}$$

$$\leq 4\,|z+s| \int_0^{+\infty} \frac{t}{|(z+s)^2+t^2|^2} \frac{dt}{e^{2\pi t}-1}$$

$$\leq \frac{4}{|z+s|^3} \int_0^{+\infty} \frac{t\,dt}{e^{2\pi t}-1} \to 0 \quad (s \to +\infty).$$

Adding $1/(2z^2)$ to both sides of (6.71) and making $s \to +\infty$ we get, by (6.72) and (6.73),

$$\sum_{k=0}^{\infty} \frac{1}{(z+k)^2} = \frac{1}{z} + \frac{1}{2z^2} + 4\int_0^{+\infty} \frac{tz}{(z^2+t^2)^2} \frac{dt}{e^{2\pi t}-1}.$$

Writing w in place of z and using (6.52) we obtain

$$\psi'(w) = \frac{1}{w} + \frac{1}{2w^2} + 4\int_0^{+\infty} \frac{w}{(w^2+t^2)^2}\,\frac{t\,\mathrm{d}t}{e^{2\pi t}-1} \qquad (\mathrm{Re}\, w > 0). \tag{6.74}$$

For $\mathrm{Re}\, z > 0$ and for any w on the segment $[1, z]$ of endpoints 1 and z we have $|w| \le \max\{1, |z|\}$. Also, let $\delta = \pi/2 - |\arg w|$, whence $0 < \delta \le \pi/2$. If $\delta \ge \pi/4$ then, for any $t > 0$, $|w^2 + t^2| \ge |w|^2 \ge (\mathrm{Re}\, w)^2$. If $\delta < \pi/4$ then $|w^2 + t^2| \ge |w|^2 \sin(2\delta) > |w|^2 \sin\delta = |w|\,\mathrm{Re}\, w \ge (\mathrm{Re}\, w)^2$. In either case

$$|w^2 + t^2| \ge (\mathrm{Re}\, w)^2 \ge (\min\{1, \mathrm{Re}\, z\})^2,$$

whence

$$\left| \frac{w}{(w^2+t^2)^2} \right| \le \frac{\max\{1, |z|\}}{(\min\{1, \mathrm{Re}\, z\})^4},$$

which shows that the integral on the right-hand side of (6.74) converges absolutely and uniformly for $w \in [1, z]$. Thus, after integrating (6.74) for $w \in [1, z]$, we may interchange the integrations. We get

$$\psi(z) = \frac{\Gamma'(z)}{\Gamma(z)} = \log z - \frac{1}{2z} + C_1 - 2\int_0^{+\infty} \frac{t\,\mathrm{d}t}{(z^2+t^2)(e^{2\pi t}-1)},$$

where C_1 is a constant. Integrating again,

$$\log \Gamma(z) = \left(z - \frac{1}{2}\right)\log z + (C_1 - 1)z + C_2 + 2\int_0^{+\infty} \frac{\arctan(t/z)}{e^{2\pi t}-1}\,\mathrm{d}t, \tag{6.75}$$

where C_2 is a constant. For $t, z > 0$ we have $0 \le \arctan(t/z) \le t/z$, whence

$$0 \le \int_0^{+\infty} \frac{\arctan(t/z)}{e^{2\pi t}-1}\,\mathrm{d}t \le \frac{1}{z}\int_0^{+\infty} \frac{t\,\mathrm{d}t}{e^{2\pi t}-1} = O\!\left(\frac{1}{z}\right).$$

Thus for $z > 0$, $z \to +\infty$, we get

$$\log \Gamma(z) = \left(z - \frac{1}{2}\right)\log z + (C_1 - 1)z + C_2 + O\!\left(\frac{1}{z}\right).$$

Comparing with (6.39) we obtain $C_1 = 0$, $C_2 = \log\sqrt{2\pi}$. The theorem follows from (6.75). □

Chapter 7
Linear Differential Equations

In this chapter we deal with some basic facts concerning ordinary linear differential equations in the analytic domain, culminating in Fuchs' theory on regular singular points. Since several special functions of interest in various applications, including the hypergeometric functions ${}_2F_1$, $\Phi = {}_1F_1$ and Ψ discussed in the next chapter, are solutions of homogeneous linear differential equations of second order, we shall give a full proof of Fuchs' theorem in the case of order $n = 2$. For a treatment of Fuchs' theory in the general n case, the reader is referred to E. L. Ince's book quoted in the bibliography.

7.1 Cauchy's Method

We expound in this section Cauchy's original method, based on the so-called dominant functions, to prove the existence, in the analytic domain, of a unique solution of an ordinary homogeneous linear differential equation of order n with initial conditions, in a disc where the coefficients of the differential equation are regular functions. For simplicity, and in view of subsequent applications, we confine ourselves to treating *linear* differential equations, but such a restriction is not essential in this section, and Cauchy's method can be adapted to more general differential equations of order n in the analytic domain.

We begin with the following definition.

Definition 7.1 Let the functions $f(z)$ and $g(z)$ be regular at $z_0 \in \mathbb{C}$. If

$$\left|f^{(k)}(z_0)\right| \leq g^{(k)}(z_0) \quad \text{for all } k = 0, 1, 2, \dots,$$

we say that $g(z)$ is dominant over $f(z)$ at z_0.

C. Viola, *An Introduction to Special Functions*,
UNITEXT - La Matematica per il 3+2 102, DOI 10.1007/978-3-319-41345-7_7

If $g(z)$ is dominant over $f(z)$ at z_0, and if the Taylor series

$$g(z) = \sum_{k=0}^{\infty} \frac{g^{(k)}(z_0)}{k!}(z - z_0)^k \tag{7.1}$$

converges absolutely at a point z, then also

$$f(z) = \sum_{k=0}^{\infty} \frac{f^{(k)}(z_0)}{k!}(z - z_0)^k \tag{7.2}$$

converges absolutely at z. Hence the radius of convergence of the series (7.1) does not exceed the radius of convergence of (7.2).

Theorem 7.1 (Cauchy) *Let $R > 0$, $z_0 \in \mathbb{C}$, and $n \in \mathbb{N}$, $n \geq 1$. Let*

$$a_1(z), \ldots, a_n(z)$$

be n functions regular in the closed disc $|z - z_0| \leq R$. Then, given n arbitrary constants $w_0, w_0', \ldots, w_0^{(n-1)} \in \mathbb{C}$, there exists a unique function $w(z)$, regular in the open disc $|z - z_0| < R$, satisfying the differential equation

$$w^{(n)} = a_1(z)\, w^{(n-1)} + \cdots + a_{n-1}(z)\, w' + a_n(z)\, w \tag{7.3}$$

and the initial conditions

$$w(z_0) = w_0,\ w'(z_0) = w_0', \ldots, w^{(n-1)}(z_0) = w_0^{(n-1)}. \tag{7.4}$$

Proof Let

$$a_\nu(z) = \sum_{\varrho=0}^{\infty} a_{\nu\varrho}(z - z_0)^\varrho \qquad (\nu = 1, \ldots, n) \tag{7.5}$$

be the Taylor expansions in the disc $|z - z_0| \leq R$ of the coefficients of the differential equation (7.3). If the Cauchy problem (7.3), (7.4) has a solution

$$w(z) = \sum_{r=0}^{\infty} c_r (z - z_0)^r \tag{7.6}$$

in a neighbourhood of z_0 then, by (7.4),

$$c_0 = w_0,\ c_1 = \frac{w_0'}{1!}, \ldots, c_{n-1} = \frac{w_0^{(n-1)}}{(n-1)!}. \tag{7.7}$$

Differentiating n times the series (7.6) and substituting in (7.3) the $n+1$ series for $w(z), w'(z), \dots, w^{(n)}(z)$ thus obtained, as well as the n series (7.5), we obtain linear recurrence formulae yielding successively $c_n, c_{n+1}, \dots$ from the values (7.7). Such formulae can be written as

$$c_r = \sum_{s=1}^{r} \alpha_{rs}\, c_{r-s} \quad (r = n,\ n+1, \dots), \tag{7.8}$$

where α_{rs} are certain linear combinations of $a_{\nu\varrho}$ with positive coefficients. Thus if a solution (7.6) of the Cauchy problem exists, it is unique. It remains to prove that the radius of convergence of the series (7.6) with the coefficients (7.7), (7.8) is $\geq R$.

Let $M_1, \dots, M_n > 0$ be any constants such that, in the disc $|z - z_0| \leq R$,

$$|a_\nu(z)| \leq M_\nu \quad (\nu = 1, \dots, n),$$

with the additional condition

$$M_1 > \frac{n}{R}. \tag{7.9}$$

From (7.5) we get, by Cauchy's integral formula,

$$a_{\nu\varrho} = \frac{a_\nu^{(\varrho)}(z_0)}{\varrho!} = \frac{1}{2\pi i} \oint_{|z-z_0|=R} \frac{a_\nu(z)}{(z-z_0)^{\varrho+1}}\, \mathrm{d}z.$$

Hence

$$|a_{\nu\varrho}| \leq \frac{M_\nu}{2\pi R^{\varrho+1}} \oint_{|z-z_0|=R} |\mathrm{d}z| = \frac{M_\nu}{R^\varrho}. \tag{7.10}$$

Define functions $A_1(z), \dots, A_n(z)$ as follows:

$$A_\nu(z) := \frac{M_\nu}{1 - \dfrac{z - z_0}{R}} = \sum_{\varrho=0}^{\infty} A_{\nu\varrho}(z - z_0)^\varrho \quad (\nu = 1, \dots, n). \tag{7.11}$$

Then

$$A_\nu(z) = M_\nu \sum_{\varrho=0}^{\infty} \left(\frac{z - z_0}{R}\right)^\varrho \quad (|z - z_0| < R),$$

whence, by (7.10),

$$|a_{\nu\varrho}| \leq \frac{M_\nu}{R^\varrho} = A_{\nu\varrho}. \tag{7.12}$$

Thus $A_\nu(z)$ is dominant over $a_\nu(z)$ at z_0 $(\nu = 1, \dots, n)$.

Consider the auxiliary differential equation

$$W^{(n)} = A_1(z)\,W^{(n-1)} + \ldots + A_{n-1}(z)\,W' + A_n(z)\,W. \tag{7.13}$$

Let

$$W(z) = \sum_{r=0}^{\infty} C_r (z - z_0)^r \tag{7.14}$$

be its (formal) solution satisfying the initial conditions

$$C_0 = |c_0|,\ C_1 = |c_1|, \ldots, C_{n-1} = |c_{n-1}|, \tag{7.15}$$

and let

$$C_r = \sum_{s=1}^{r} \beta_{rs} C_{r-s} \qquad (r = n,\, n+1, \ldots) \tag{7.16}$$

be the recurrence formulae analogous with (7.8), where β_{rs} are linear combinations of $A_{\nu\varrho}$ with the same positive coefficients as in α_{rs}. Thus, by (7.12),

$$|\alpha_{rs}| \le \beta_{rs} \quad \text{with } \beta_{rs} > 0. \tag{7.17}$$

From (7.8), (7.15), (7.16) and (7.17) we get, by induction on r,

$$|c_r| \le C_r \qquad (r = 0, 1, 2, \ldots). \tag{7.18}$$

In particular

$$\sum_{r=0}^{\infty} |c_r (z - z_0)^r| \le \sum_{r=0}^{\infty} C_r |z - z_0|^r.$$

We will prove that, apart from the case $c_0 = c_1 = \cdots = c_{n-1} = 0$ which yields $W(z) \equiv 0$ by (7.15) and (7.16), the radius of convergence of the series (7.14) is R. Then, by (7.18), $W(z)$ is dominant over $w(z)$ at z_0, whence the radius of convergence of the series (7.6) is $\ge R$, as desired.

We make the change of variable $z - z_0 = RZ$. Then, for $\nu = 1, \ldots, n$,

$$\frac{\mathrm{d}^\nu W}{\mathrm{d}Z^\nu} = R^\nu \frac{\mathrm{d}^\nu W}{\mathrm{d}z^\nu}$$

and, by (7.11),

$$A_\nu(z) = \frac{M_\nu}{1 - Z}.$$

Thus (7.13) becomes

$$(1 - Z)\,\frac{\mathrm{d}^n W}{\mathrm{d}Z^n} = M_1 R\,\frac{\mathrm{d}^{n-1} W}{\mathrm{d}Z^{n-1}} + \ldots + M_{n-1} R^{n-1}\,\frac{\mathrm{d}W}{\mathrm{d}Z} + M_n R^n W. \tag{7.19}$$

The series (7.14) is

$$W(Z) = \sum_{r=0}^{\infty} \gamma_r Z^r \tag{7.20}$$

with

$$\gamma_r = C_r R^r \geq 0. \tag{7.21}$$

Since we have excluded the trivial case $c_0 = c_1 = \cdots = c_{n-1} = 0$, from (7.15), (7.16) and (7.17) we get

$$\gamma_r = C_r R^r > 0 \quad \text{for} \quad r \geq n. \tag{7.22}$$

Substituting (7.20) into (7.19) we easily obtain the recursion

$$r!\,\gamma_r = (r-n)(r-1)!\,\gamma_{r-1} + \sum_{s=1}^{n} (r-s)!\,M_s R^s \gamma_{r-s} \quad (r = n,\, n+1,\, \ldots). \tag{7.23}$$

By (7.21) and (7.23),

$$\begin{aligned} \gamma_r &= \frac{r - n + M_1 R}{r}\,\gamma_{r-1} + \sum_{s=2}^{n} \frac{(r-s)!}{r!}\,M_s R^s \gamma_{r-s} \\ &\geq \frac{r - n + M_1 R}{r}\,\gamma_{r-1} \quad (r \geq n) \end{aligned} \tag{7.24}$$

whence, by (7.9) and (7.22),

$$0 < \gamma_n < \gamma_{n+1} < \gamma_{n+2} < \ldots \tag{7.25}$$

Hence from (7.24) we get, for $r \geq n+1$,

$$\frac{\gamma_r}{\gamma_{r-1}} = \frac{r - n + M_1 R}{r} + \sum_{s=2}^{n} \frac{M_s R^s}{(r-s+1)\cdots(r-1)\,r}\,\frac{\gamma_{r-s}}{\gamma_{r-1}}. \tag{7.26}$$

By (7.25), $\gamma_{r-s}/\gamma_{r-1} < 1$ for $r \geq 2n$ and $s = 2, \ldots, n$. Thus, by (7.26),

$$\lim_{r\to\infty} \frac{\gamma_r}{\gamma_{r-1}} = 1.$$

Therefore the series (7.20) has radius of convergence 1, whence (7.14) has radius of convergence R, as claimed. □

7.2 Singular Points of the Coefficients

If the coefficients $a_\nu(z)$ of (7.3) are regular functions in a neighbourhood of a point $z_0 \in \mathbb{C}$ except possibly at z_0, one generally expects the solutions w of the differential equation (7.3) to have a branch point at z_0, as the following example shows. Let $a(z)$ have a simple pole at $z = 0$ with residue r:

$$a(z) = \frac{r}{z} + b(z),$$

where $b(z)$ is regular in a disc $|z| < R$, and assume that the constant r is not an integer:

$$r = \operatorname*{Res}_{z=0} a(z) \in \mathbb{C} \setminus \mathbb{Z}.$$

Let $z_1 \in \mathbb{C}$ be any fixed point with $0 < |z_1| < R$. The solution

$$w(z) = \exp \int_{z_1}^{z} a(t)\,\mathrm{d}t \tag{7.27}$$

of the first order equation

$$w' = a(z)w$$

has a branch point at $z = 0$, because moving z, inside the punctured disc $0 < |z| < R$, along a circuit beginning and ending at z_1 and enclosing 0 changes the value of (7.27) from 1 to $e^{2\pi i r} \neq 1$.

We discuss in detail the case of a second order homogeneous linear differential equation. We write the equation in the form

$$w'' + p_1(z)\,w' + p_2(z)\,w = 0, \tag{7.28}$$

and assume that the functions $p_1(z)$ and $p_2(z)$ are one-valued and regular in a punctured disc

$$\mathcal{P} = \{z \in \mathbb{C} \,\big|\, 0 < |z - z_0| < R\}, \tag{7.29}$$

where z_0 is a singular point (i.e., either a pole or an isolated essential singularity) for at least one of $p_1(z)$ and $p_2(z)$.

Let $\mathcal{D}_1 = \{z \in \mathbb{C} \,\big|\, |z - z_1| < \varrho_1\}$ and $\mathcal{D}_2 = \{z \in \mathbb{C} \,\big|\, |z - z_2| < \varrho_2\}$ be two overlapping discs contained in $\mathcal{P}$, and, by Theorem 7.1, let $w_1(z)$ be any solution of (7.28) in $\mathcal{D}_1$. We take a point $z_3 \in \mathcal{D}_1 \cap \mathcal{D}_2$, lying on the segment of endpoints z_1, z_2

and satisfying $\varrho_1 - |z_1 - z_3| < \varrho_2 - |z_2 - z_3|$, i.e., closer to $\partial\mathcal{D}_1$ than to $\partial\mathcal{D}_2$. We can apply Theorem 7.1 to the differential equation (7.28) in the disc

$$\mathcal{D}_3 = \{z \in \mathbb{C} \mid |z - z_3| < \varrho_3\}, \quad \varrho_3 = \varrho_2 - |z_2 - z_3|,$$

with initial conditions $w(z_3) = w_1(z_3)$ and $w'(z_3) = w_1'(z_3)$. Thus $w_1(z)$ can be analytically continued to $\mathcal{D}_1 \cup \mathcal{D}_3$. Then we take a point $z_4 \in \mathcal{D}_3 \setminus \mathcal{D}_1$ lying on the segment of endpoints z_2, z_3, and apply Theorem 7.1 to (7.28) in the disc

$$\mathcal{D}_4 = \{z \in \mathbb{C} \mid |z - z_4| < \varrho_4\}, \quad \varrho_4 = \varrho_2 - |z_2 - z_4|,$$

with initial conditions $w(z_4) = w_1(z_4)$ and $w'(z_4) = w_1'(z_4)$, so that $w_1(z)$ is analytically continued to $\mathcal{D}_1 \cup \mathcal{D}_4$. Iterating this process finitely many times, we can analytically continue the solution $w_1(z)$ of (7.28) to $\mathcal{D}_1 \cup \mathcal{D}_2$. This shows that for any simply connected open set $\Omega \subset \mathcal{P}$ such that $z_1 \in \Omega$, $w_1(z)$ can be analytically continued to the whole Ω.

Now, starting from any point $z \in \Omega$, we move in the positive sense along a simple closed circuit $\Gamma \subset \mathcal{P}$ enclosing z_0. By compactness of Γ, we can apply Theorem 7.1 to the equation (7.28) finitely many times, using successive overlapping discs contained in $\mathcal{P}$ with centres lying on Γ. Let $W_1(z)$ be the value thus obtained by analytic continuation from $w_1(z)$, after returning along Γ to the initial point $z \in \Omega$. Since the differential equation (7.28) is identically satisfied along Γ, $W_1(z)$ is a solution of (7.28) in Ω. Similarly, let $w_2(z)$ be another solution of (7.28) in Ω, and let $W_2(z)$ be the solution of (7.28) in Ω obtained from $w_2(z)$ by analytic continuation along Γ. If $w_1(z)$ and $w_2(z)$ are linearly independent, there exist constants h_{11}, h_{12}, h_{21}, h_{22}, depending on $w_1(z)$ and $w_2(z)$ but not on Γ, such that for any $z \in \Omega$

$$\begin{cases} W_1(z) = h_{11}\, w_1(z) + h_{12}\, w_2(z) \\ W_2(z) = h_{21}\, w_1(z) + h_{22}\, w_2(z). \end{cases} \tag{7.30}$$

Plainly

$$\begin{vmatrix} h_{11} & h_{12} \\ h_{21} & h_{22} \end{vmatrix} \neq 0, \tag{7.31}$$

for otherwise $W_1(z)$ and $W_2(z)$ would be linearly dependent. Then there would exist constants μ_1, μ_2, not both zero, such that $\mu_1 W_1(z) + \mu_2 W_2(z) = 0$ identically in Ω, whence, moving along Γ in the negative sense, one would get by analytic continuation $\mu_1 w_1(z) + \mu_2 w_2(z) = 0$ identically in Ω, contradicting the linear independence of $w_1(z)$ and $w_2(z)$.

Let $w(z)$ be any not identically vanishing solution of (7.28) in Ω, and let $k_1, k_2 \in \mathbb{C}$ be such that

$$w(z) = k_1 w_1(z) + k_2 w_2(z).$$

If $W(z)$ is the solution of (7.28) in Ω obtained from $w(z)$ by analytic continuation after describing once the circuit Γ in the positive sense, we get

$$W(z) = k_1 W_1(z) + k_2 W_2(z),$$

whence, by (7.30),

$$W(z) = (k_1 h_{11} + k_2 h_{21}) w_1(z) + (k_1 h_{12} + k_2 h_{22}) w_2(z).$$

Thus the condition for $w(z)$ to be invariant, up to a multiplicative constant λ, after describing the circuit Γ, i.e. $W(z) = \lambda w(z)$, is that λ satisfies $k_1 h_{11} + k_2 h_{21} = \lambda k_1$ and $k_1 h_{12} + k_2 h_{22} = \lambda k_2$, i.e.,

$$\begin{cases} (h_{11} - \lambda) k_1 + h_{21} k_2 = 0 \\ h_{12} k_1 + (h_{22} - \lambda) k_2 = 0. \end{cases} \tag{7.32}$$

This linear system is satisfied by suitable $(k_1, k_2) \neq (0, 0)$ if and only if

$$\begin{vmatrix} h_{11} - \lambda & h_{21} \\ h_{12} & h_{22} - \lambda \end{vmatrix} = 0, \tag{7.33}$$

i.e., if and only if λ is an eigenvalue of the matrix

$$\begin{pmatrix} h_{11} & h_{12} \\ h_{21} & h_{22} \end{pmatrix}.$$

The equation (7.33) is called the *characteristic equation* of the singular point z_0. Although h_{11}, h_{12}, h_{21}, h_{22} depend on the fundamental system $\big(w_1(z), w_2(z)\big)$ of solutions of (7.28) in Ω, it is easily seen that the roots λ_1, λ_2 of the characteristic equation (7.33) depend on z_0 but are independent of $\big(w_1(z), w_2(z)\big)$. Indeed, the non-zero solutions $w(z)$ of (7.28) which are invariant, up to a multiplicative constant λ, after describing once in the positive sense a circuit Γ enclosing z_0 are obviously independent of the fundamental system $\big(w_1(z), w_2(z)\big)$, and so are the corresponding quotients $\lambda = W(z)/w(z)$.

By (7.31), the roots λ_1, λ_2 of (7.33) satisfy $\lambda_1 \neq 0$ and $\lambda_2 \neq 0$. Taking $\lambda = \lambda_1$ in (7.32), we can find a solution $(k_1, k_2) \neq (0, 0)$ of (7.32), which is unique up to a constant factor unless the rank of (7.33) for $\lambda = \lambda_1$ is zero, i.e., unless $h_{11} = h_{22} = \lambda_1$, $h_{12} = h_{21} = 0$. In this case we get $\lambda_1 = \lambda_2$ and, by (7.30), all the solutions of (7.28) in Ω, after describing Γ, are transformed by multiplication by λ_1. In either case, whether the rank of (7.33) for $\lambda = \lambda_1$ is 1 or 0, let $(k_1, k_2) \neq (0, 0)$ be a solution of (7.32) for $\lambda = \lambda_1$. Let

$$w_1^*(z) = k_1 w_1(z) + k_2 w_2(z)$$

be the corresponding solution of (7.28) invariant up to multiplication by λ_1, and

$$W_1^*(z) = k_1 W_1(z) + k_2 W_2(z) = \lambda_1 w_1^*(z).$$

Define

$$r_1 = \frac{1}{2\pi i} \log \lambda_1. \tag{7.34}$$

Since the difference of any two logarithms of λ_1 is $2l\pi i$ with $l \in \mathbb{Z}$, r_1 is defined modulo 1. Let $z - z_0 = \varrho e^{i\vartheta}$ ($\varrho > 0$, $\vartheta \in \mathbb{R}$). Then the function $(z - z_0)^{r_1} = \varrho^{r_1} e^{ir_1\vartheta}$ is multiplied by $e^{2\pi i r_1} = \lambda_1$ after describing once in the positive sense the circuit Γ. Since the functions $w_1^*(z)$ and $(z - z_0)^{r_1}$ are both multiplied by λ_1, the quotient $\varphi_1(z) := w_1^*(z)/(z - z_0)^{r_1}$ is unchanged after describing Γ. Therefore the differential equation (7.28) has the solution

$$w_1^*(z) = (z - z_0)^{r_1} \varphi_1(z), \tag{7.35}$$

where the function $\varphi_1(z)$ is one-valued, regular and not identically zero in the punctured disc $\mathcal{P}$ defined by (7.29).

We distinguish two cases.

First case: $\lambda_1 \neq \lambda_2$. Let

$$r_2 = \frac{1}{2\pi i} \log \lambda_2. \tag{7.36}$$

We repeat the above discussion with λ_2 in place of λ_1, and we conclude that (7.28) has the solutions

$$w_1^*(z) = (z - z_0)^{r_1} \varphi_1(z) \quad \text{and} \quad w_2^*(z) = (z - z_0)^{r_2} \varphi_2(z) \tag{7.37}$$

with r_1 and r_2, given by (7.34) and (7.36), defined modulo 1, $r_1 - r_2 \notin \mathbb{Z}$, and with $\varphi_1(z)$ and $\varphi_2(z)$ one-valued, regular and not identically zero in $\mathcal{P}$. Plainly the two solutions (7.37) are linearly independent, for otherwise one would get $(z - z_0)^{r_1} \varphi_1(z) = C(z - z_0)^{r_2} \varphi_2(z)$ with a constant $C \neq 0$, so that, after describing the circuit Γ,

$$\lambda_1 (z - z_0)^{r_1} \varphi_1(z) = \lambda_1 C (z - z_0)^{r_2} \varphi_2(z) = C \lambda_2 (z - z_0)^{r_2} \varphi_2(z),$$

whence $\lambda_1 = \lambda_2$. Therefore, if $\lambda_1 \neq \lambda_2$ the two functions (7.37) form a fundamental system of solutions of (7.28), and all the solutions of (7.28) in $\mathcal{P}$ are given by

$$w(z) = C_1 (z - z_0)^{r_1} \varphi_1(z) + C_2 (z - z_0)^{r_2} \varphi_2(z),$$

with arbitrary constants C_1 and C_2.

Second case: $\lambda_1 = \lambda_2$. We consider the fundamental system of solutions of (7.28) obtained by associating with the solution $w_1^*(z)$ given by (7.35) any other solution

$\widetilde{w}_2(z)$, regular in Ω, such that $w_1^*(z)$ and $\widetilde{w}_2(z)$ are linearly independent. If $W_1^*(z)$ and $\widetilde{W}_2(z)$ are the solutions of (7.28) obtained by analytic continuation from $w_1^*(z)$ and $\widetilde{w}_2(z)$ respectively after describing the circuit Γ, (7.30) becomes

$$\begin{cases} W_1^*(z) = \lambda_1 w_1^*(z) \\ \widetilde{W}_2(z) = h_{21} w_1^*(z) + h_{22}\widetilde{w}_2(z). \end{cases}$$

Thus the characteristic equation (7.33) is

$$\begin{vmatrix} \lambda_1 - \lambda & h_{21} \\ 0 & h_{22} - \lambda \end{vmatrix} = 0, \tag{7.38}$$

i.e., $(\lambda_1 - \lambda)(h_{22} - \lambda) = 0$; and since $\lambda_1 = \lambda_2$ is a double root of (7.38), we get $h_{22} = \lambda_1 = \lambda_2$. Hence (7.30) is now

$$\begin{cases} W_1^*(z) = \lambda_1 w_1^*(z) \\ \widetilde{W}_2(z) = h_{21} w_1^*(z) + \lambda_1\widetilde{w}_2(z). \end{cases}$$

Therefore

$$\frac{\widetilde{W}_2(z)}{W_1^*(z)} = \frac{\widetilde{w}_2(z)}{w_1^*(z)} + \frac{h_{21}}{\lambda_1}.$$

On describing the circuit Γ, $\arg(z - z_0)$ increases by 2π, whence the function

$$\frac{h_{21}}{2\pi i\lambda_1}\log(z - z_0) = \frac{h_{21}}{2\pi i\lambda_1}\log|z - z_0| + \frac{h_{21}}{2\pi\lambda_1}\arg(z - z_0)$$

changes to

$$\frac{h_{21}}{2\pi i\lambda_1}\log(z - z_0) + \frac{h_{21}}{\lambda_1}.$$

Thus, defining the constant

$$H := \frac{h_{21}}{2\pi i\lambda_1}, \tag{7.39}$$

we see that the function

$$\psi_2(z) := \frac{\widetilde{w}_2(z)}{w_1^*(z)} - H\log(z - z_0)$$

is unchanged after describing Γ. We conclude that, in the present case $\lambda_1 = \lambda_2$, (7.28) has the two linearly independent solutions

$w_1^*(z) = (z - z_0)^{r_1}\varphi_1(z)$ and

$$\widetilde{w}_2(z) = (z - z_0)^{r_1}\varphi_1(z)\big(H\log(z - z_0) + \psi_2(z)\big), \quad (7.40)$$

where r_1 is given by (7.34), $\varphi_1(z)$ is one-valued, regular and not identically zero in the punctured disc $\mathcal{P}$, and $\psi_2(z)$ is one-valued and meromorphic in $\mathcal{P}$. Thus all the solutions of (7.28) in $\mathcal{P}$ are given by

$$w(z) = (z - z_0)^{r_1}\varphi_1(z)\big(C_1 + C_2\big(H\log(z - z_0) + \psi_2(z)\big)\big)$$

with arbitrary constants C_1 and C_2. We remark that $\psi_2(z)$ can be defined up to an arbitrary additive constant C, because replacing in (7.40) $\psi_2(z)$ by $\psi_2(z) + C$ changes $\widetilde{w}_2(z)$ to $\widetilde{w}_2(z) + Cw_1^*(z)$, with $w_1^*(z)$ and $\widetilde{w}_2(z) + Cw_1^*(z)$ linearly independent. Moreover, if $H \neq 0$ the second solution $\widetilde{w}_2(z)$ in (7.40) can be written as

$$(z - z_0)^{r_1}(H\varphi_1(z))\Big(\log(z - z_0) + \frac{\psi_2(z)}{H}\Big),$$

so that, by changing $H\varphi_1(z)$ to $\varphi_1(z)$ and $\psi_2(z)/H$ to $\psi_2(z)$, one may assume $H = 1$ in (7.40). This is the so-called logarithmic case, corresponding to (7.38) with $h_{22} = \lambda_1$ and with rank 1 for $\lambda = \lambda_1$ since, by (7.39), $h_{21} = 2\pi i\lambda_1 H \neq 0$. If $H = 0$, i.e. if $h_{21} = 0$, the characteristic equation (7.38) is

$$\begin{vmatrix} \lambda_1 - \lambda & 0 \\ 0 & \lambda_1 - \lambda \end{vmatrix} = 0.$$

Then the linear independence of $w_1^*(z)$ and $\widetilde{w}_2(z)$ implies that $\psi_2(z)$ is not constant, and we are in the case already discussed of (7.33) with rank zero for $\lambda = \lambda_1$.

With the above discussion we have proved the following

Theorem 7.2 *Let the functions $p_1(z)$ and $p_2(z)$ be one-valued and regular in the punctured disc*

$$\mathcal{P} = \{z \in \mathbb{C} \,|\, 0 < |z - z_0| < R\},$$

where z_0 is a singular point for at least one of $p_1(z)$ and $p_2(z)$. Then the differential equation

$$w'' + p_1(z)\,w' + p_2(z)\,w = 0 \quad (7.41)$$

has in $\mathcal{P}$ two linearly independent solutions $w_1(z)$ and $w_2(z)$ of the form

$$w_1(z) = (z - z_0)^{r_1}\varphi_1(z) \quad \textit{and} \quad w_2(z) = (z - z_0)^{r_2}\varphi_2(z) \quad (7.42)$$

with $r_1, r_2 \in \mathbb{C}$, $r_1 - r_2 \notin \mathbb{Z}$, given by (7.34) *and* (7.36) *and with $\varphi_1(z)$ and $\varphi_2(z)$ one-valued and regular in $\mathcal{P}$ if the roots λ_1, λ_2 of the characteristic equation* (7.33) *are distinct, or of the form*

$$w_1(z) = (z-z_0)^{r_1}\varphi_1(z) \quad and$$
$$w_2(z) = (z-z_0)^{r_1}\varphi_1(z)\big(H\log(z-z_0)+\psi_2(z)\big) \quad (7.43)$$

with $r_1 \in \mathbb{C}$ given by (7.34), *$H \in \mathbb{C}$ given by* (7.39), *$\varphi_1(z)$ one-valued and regular in $\mathcal{P}$ and $\psi_2(z)$ one-valued and meromorphic in $\mathcal{P}$ if the characteristic equation* (7.33) *has the double root λ_1.*

7.3 Fuchs' Theorem

If no additional conditions are assumed on the nature of the singularity z_0 for the coefficients $p_1(z)$ and $p_2(z)$ of the differential equation (7.41), one generally expects z_0 to be an isolated essential singularity for the functions $\varphi_1(z)$ and $\varphi_2(z)$ in (7.42) or (7.43). In particular, by Picard's second theorem (Theorem 1.4), $\varphi_1(z)$ has in general infinitely many zeros in any neighbourhood of z_0. Thus if the characteristic equation (7.33) has the double root λ_1, by (7.43) $\psi_2(z) = w_2(z)/w_1(z) - H\log(z-z_0)$ is generally expected to have infinitely many poles in any neighbourhood of z_0, and therefore not even to possess a Laurent series expansion at z_0. However, in most cases of differential equations (7.41) arising from applications, the functions $\varphi_1(z)$ and $\varphi_2(z)$, or $\varphi_1(z)$ and $\psi_2(z)$, turn out to have a pole, or even to be regular, at z_0. We incidentally remark that if $\varphi_1(z)$ has a pole or a zero at z_0, i.e., if $\varphi_1(z) = (z-z_0)^N\Phi_1(z)$ with $N \in \mathbb{Z}$ and with $\Phi_1(z)$ regular and $\neq 0$ at z_0, then $(z-z_0)^{r_1}\varphi_1(z) = (z-z_0)^{r_1+N}\Phi_1(z)$. Since r_1 is defined modulo 1, one can replace r_1 with r_1+N and hence $\varphi_1(z)$ with $\Phi_1(z)$. Thus in this case one can assume $\varphi_1(z)$ to be regular and $\neq 0$ at z_0, and similarly for $\varphi_2(z)$.

Therefore, a natural question to ask is what conditions can be assumed on the behaviour of $p_1(z)$ and $p_2(z)$ at z_0 to ensure that z_0 is a pole or a regular point for $\varphi_1(z)$ and $\varphi_2(z)$, or for $\varphi_1(z)$ and $\psi_2(z)$. An answer to this question is given by the following

Theorem 7.3 (Fuchs) *Under the assumptions of Theorem 7.2, the functions $\varphi_1(z)$, $\varphi_2(z)$ and $\psi_2(z)$ in* (7.42) *and* (7.43) *have at most a pole at z_0, if and only if $p_1(z)$ has at most a simple pole and $p_2(z)$ at most a double pole at z_0.*

If such conditions are satisfied, z_0 is said to be a regular singular point, or a fuchsian singular point, of the differential equation (7.41).

Proof First we prove that Fuchs' conditions on $p_1(z)$ and $p_2(z)$ are necessary. We assume that (7.41) has the fundamental system of solutions (7.42) or (7.43), where $\varphi_1(z)$ and $\varphi_2(z)$, or $\varphi_1(z)$ and $\psi_2(z)$, either are regular or have a pole (of any order) at z_0. We can easily express the coefficients $p_1(z)$ and $p_2(z)$ of (7.41) through the solutions $w_1(z)$ and $w_2(z)$. From the identities

$$w_1'' + p_1(z)w_1' + p_2(z)w_1 = 0 \quad \text{and} \quad w_2'' + p_1(z)w_2' + p_2(z)w_2 = 0, \qquad (7.44)$$

multiplying the first by $-w_2$, the second by w_1 and summing, we obtain

$$w_1 w_2'' - w_2 w_1'' + p_1(z)(w_1 w_2' - w_2 w_1') = 0.$$

Dividing by the wronskian $w_1 w_2' - w_2 w_1'$, which is $\neq 0$, we get

$$\begin{aligned} p_1(z) &= -\frac{w_1 w_2'' - w_2 w_1''}{w_1 w_2' - w_2 w_1'} = -\frac{\mathrm{d}}{\mathrm{d}z} \log(w_1 w_2' - w_2 w_1') \\ &= -\frac{\mathrm{d}}{\mathrm{d}z} \log\left(w_1^2 \frac{\mathrm{d}}{\mathrm{d}z}\left(\frac{w_2}{w_1}\right)\right), \end{aligned} \tag{7.45}$$

and, by the first of (7.44),

$$p_2(z) = -\frac{w_1''}{w_1} - p_1(z)\frac{w_1'}{w_1}. \tag{7.46}$$

Let $\mathcal{F}$ be the set of functions $F(z)$ satisfying

$$F(z) = (z - z_0)^\alpha G(z),$$

where $\alpha \in \mathbb{C}$ is any constant and $G(z)$ is any function meromorphic in the disc $|z - z_0| < R$. Then the following properties hold:

(i) Products, quotients and derivatives of functions in $\mathcal{F}$ belong to $\mathcal{F}$, since

$$F'(z) = (z - z_0)^{\alpha-1}\big(\alpha G(z) + (z - z_0)G'(z)\big).$$

(ii) If $F(z) \in \mathcal{F}$ then $F'(z)/F(z)$ has at most a simple pole at z_0. For, let $G(z) = (z - z_0)^n G_0(z)$ with $n \in \mathbb{Z}$ and $G_0(z)$ regular and $\neq 0$ at z_0. Then $F(z) = (z - z_0)^{\alpha+n} G_0(z)$, whence

$$F'(z) = (\alpha + n)(z - z_0)^{\alpha+n-1} G_0(z) + (z - z_0)^{\alpha+n} G_0'(z), \tag{7.47}$$

and therefore

$$\frac{F'(z)}{F(z)} = \frac{\alpha + n}{z - z_0} + \frac{G_0'(z)}{G_0(z)}.$$

(iii) If $F(z) \in \mathcal{F}$ then $F''(z)/F(z)$ has at most a double pole at z_0. For, by (7.47),

$$\begin{aligned} F''(z) &= (\alpha + n)(\alpha + n - 1)(z - z_0)^{\alpha+n-2} G_0(z) \\ &\quad + 2(\alpha + n)(z - z_0)^{\alpha+n-1} G_0'(z) + (z - z_0)^{\alpha+n} G_0''(z), \end{aligned}$$

whence

$$\frac{F''(z)}{F(z)} = \frac{(\alpha+n)(\alpha+n-1)}{(z-z_0)^2} + \frac{2(\alpha+n)}{z-z_0}\frac{G_0'(z)}{G_0(z)} + \frac{G_0''(z)}{G_0(z)}.$$

If (7.41) has the fundamental system of solutions (7.42), then $w_1, w_2 \in \mathcal{F}$ whence, by the above property (i), $w_1^2 \dfrac{\mathrm{d}}{\mathrm{d}z}\left(\dfrac{w_2}{w_1}\right) \in \mathcal{F}$. By (7.45), $-p_1(z)$ is the logarithmic derivative of a function belonging to $\mathcal{F}$ and therefore, by (ii), has at most a simple pole at z_0. Also, again by (ii), w_1'/w_1 has at most a simple pole at z_0, whence $p_1(z)\, w_1'/w_1$ has at most a double pole. By (iii) w_1''/w_1 has at most a double pole, whence, by (7.46), $p_2(z)$ has at most a double pole at z_0.

If (7.41) has the fundamental system of solutions (7.43), then $w_1 \in \mathcal{F}$ and, by (7.45), $-p_1(z)$ is the logarithmic derivative of

$$w_1^2 \frac{\mathrm{d}}{\mathrm{d}z}\big(H \log(z-z_0) + \psi_2(z)\big) = w_1^2 (z-z_0)^{-1}\big(H + (z-z_0)\psi_2'(z)\big)$$

which belongs to $\mathcal{F}$ by property (i). By (ii), $-p_1(z)$ has at most a simple pole at z_0. Moreover, as in the previous case, by applying (7.46) we conclude that $p_2(z)$ has at most a double pole at z_0.

Next we prove that Fuchs' conditions on $p_1(z)$ and $p_2(z)$ are sufficient. We do this through a constructive method, due to Frobenius, which allows us to determine the exponents r_1 and r_2 in (7.42) or (7.43) so that, according to a remark above, $\varphi_1(z)$ and $\varphi_2(z)$ are regular and $\neq 0$ at z_0, and also to find the Taylor expansions of $\varphi_1(z)$ and $\varphi_2(z)$ and the Laurent expansion of $\psi_2(z)$ at z_0.

By Fuchs' conditions we have

$$p_1(z) = \frac{A(z)}{z-z_0} \quad \text{and} \quad p_2(z) = \frac{B(z)}{(z-z_0)^2}, \tag{7.48}$$

where $A(z)$ and $B(z)$ are regular at z_0 and therefore are sums of Taylor series

$$A(z) = \sum_{n=0}^{\infty} a_n (z-z_0)^n, \quad B(z) = \sum_{n=0}^{\infty} b_n (z-z_0)^n \qquad (|z-z_0| < R), \tag{7.49}$$

with R as in the statement of Theorem 7.2. Hence the differential equation (7.41) can be written as

$$(z-z_0)^2 w'' + A(z)(z-z_0)w' + B(z)w = 0. \tag{7.50}$$

We seek solutions of (7.50) of the form

$$w(z) = (z - z_0)^r \sum_{n=0}^{\infty} c_n (z - z_0)^n, \tag{7.51}$$

with complex constants r and c_n to be determined, where c_0 is an arbitrary constant satisfying

$$c_0 \neq 0. \tag{7.52}$$

From (7.51) we get

$$(z - z_0)w' = (z - z_0)^r \sum_{n=0}^{\infty} (r + n)c_n (z - z_0)^n$$

and

$$(z - z_0)^2 w'' = (z - z_0)^r \sum_{n=0}^{\infty} (r + n)(r + n - 1)c_n (z - z_0)^n.$$

Substituting in (7.50) and dividing by the common factor $(z - z_0)^r$ we obtain, by (7.49),

$$\sum_{n=0}^{\infty} (r + n)(r + n - 1)c_n (z - z_0)^n + \left(\sum_{n=0}^{\infty} (r + n)c_n (z - z_0)^n\right)\left(\sum_{n=0}^{\infty} a_n (z - z_0)^n\right)$$
$$+ \left(\sum_{n=0}^{\infty} c_n (z - z_0)^n\right)\left(\sum_{n=0}^{\infty} b_n (z - z_0)^n\right) = 0,$$

i.e.,

$$\sum_{n=0}^{\infty} \left((r + n)(r + n - 1)c_n + \sum_{m=0}^{n} \big((r + m)a_{n-m} + b_{n-m}\big)c_m\right)(z - z_0)^n = 0.$$

Hence the condition for (7.51) to satisfy the differential equation (7.50) is

$$\big((r + n)(r + n - 1) + (r + n)a_0 + b_0\big)c_n$$
$$= -\sum_{m=0}^{n-1} \big((r + m)a_{n-m} + b_{n-m}\big)c_m \quad (n = 0, 1, 2, \dots), \tag{7.53}$$

provided the Taylor series $\sum c_n (z - z_0)^n$ with the coefficients c_n obtained from (7.53) has radius of convergence > 0.

We can conveniently write the recurrence formula (7.53) by making use of auxiliary functions $f_\nu(t)$ which we define as follows:

$$\begin{aligned} f_0(t) &= t(t-1) + a_0 t + b_0, \\ f_\nu(t) &= a_\nu t + b_\nu \qquad (\nu = 1, 2, 3, \ldots). \end{aligned} \tag{7.54}$$

Then (7.53) becomes

$$f_0(r+n)c_n = -\sum_{m=0}^{n-1} f_{n-m}(r+m)c_m \qquad (n = 0, 1, 2, \ldots). \tag{7.55}$$

For $n = 0$ this yields $f_0(r)c_0 = 0$ whence, by (7.52),

$$f_0(r) = r(r-1) + a_0 r + b_0 = 0. \tag{7.56}$$

Since, by (7.49), $a_0 = A(z_0)$ and $b_0 = B(z_0)$, (7.56) can be written as

$$r^2 + \big(A(z_0) - 1\big)r + B(z_0) = 0. \tag{7.57}$$

Thus the exponent r in (7.51) must be a root of (7.57), which is called the *indicial equation* of the regular singular point z_0.

We denote the roots of (7.57) by r_1 and r_2 with

$$\operatorname{Re} r_1 \geq \operatorname{Re} r_2, \tag{7.58}$$

and we set

$$s = r_1 - r_2, \tag{7.59}$$

whence $\operatorname{Re} s \geq 0$. If s is not an integer, then $\lambda_1 = e^{2\pi i r_1} \neq \lambda_2 = e^{2\pi i r_2}$, and according to the discussion in Sect. 7.2 we expect to find through the recursion (7.55) the coefficients c_n in (7.51) yielding two linearly independent solutions $w_1(z)$ and $w_2(z)$ of (7.50), corresponding to $r = r_1$ and $r = r_2$ respectively. If s is an integer ≥ 0 then $\lambda_1 = \lambda_2$, and through (7.55) we shall find only one solution $w_1(z)$ of the form (7.51), corresponding to $r = r_1$.

Taking $r = r_1$ in (7.55), we seek coefficients $c_n^{(1)}$ such that

$$f_0(r_1+n)c_n^{(1)} = -\sum_{m=0}^{n-1} f_{n-m}(r_1+m)c_m^{(1)} \qquad (n = 1, 2, 3, \ldots). \tag{7.60}$$

Since $f_0(t) = (t - r_1)(t - r_2)$, from (7.59) we get

$$f_0(r_1+n) = n(n+s) \neq 0 \qquad (n = 1, 2, 3, \ldots) \tag{7.61}$$

because $\operatorname{Re} s \geq 0$ yields $s \neq -n$. Hence the recurrence formula (7.60) yields successively $c_1^{(1)}, c_2^{(1)}, c_3^{(1)}, \ldots$ from $c_0^{(1)} \neq 0$ arbitrarily chosen. We claim that the radius of convergence of the Taylor series

$$\sum_{n=0}^{\infty} c_n^{(1)}(z-z_0)^n \tag{7.62}$$

thus obtained is $\geq R$, where R is as in (7.49).

Take any ϱ with $0 < \varrho < R$, and let $M > 0$ be a constant such that

$$\max_{|z-z_0|\leq\varrho} |A(z)| \leq M \quad \text{and} \quad \max_{|z-z_0|\leq\varrho} |B(z)| \leq M.$$

By Cauchy's integral formula

$$a_n = \frac{A^{(n)}(z_0)}{n!} = \frac{1}{2\pi i} \oint\limits_{|z-z_0|=\varrho} \frac{A(z)}{(z-z_0)^{n+1}}\,\mathrm{d}z$$

we get

$$|a_n| \leq \frac{M}{2\pi\varrho^{n+1}} \oint\limits_{|z-z_0|=\varrho} |\mathrm{d}z| = \frac{M}{\varrho^n},$$

and similarly

$$|b_n| \leq \frac{M}{\varrho^n}.$$

Therefore, by (7.54),

$$|f_\nu(t)| \leq |a_\nu|\,|t| + |b_\nu| \leq \frac{M}{\varrho^\nu}(|t|+1) \quad (\nu = 1, 2, 3, \ldots). \tag{7.63}$$

From (7.61) we get

$$|f_0(r_1+n)| = n\,|n+s| \geq n(n+\operatorname{Re} s) \geq n^2. \tag{7.64}$$

Thus, by (7.60), (7.63) and (7.64),

$$n^2|c_n^{(1)}| \leq |f_0(r_1+n)|\,|c_n^{(1)}| \leq \sum_{m=0}^{n-1} |f_{n-m}(r_1+m)|\,|c_m^{(1)}|$$

$$\leq M \sum_{m=0}^{n-1} \frac{|r_1|+m+1}{\varrho^{n-m}} |c_m^{(1)}|,$$

whence, for $n \geq 1$,

$$\left|c_n^{(1)}\right| \le \frac{M}{n}\sum_{m=0}^{n-1}\frac{|r_1|+m+1}{n}\frac{\left|c_m^{(1)}\right|}{\varrho^{n-m}} \le \frac{M}{n}\left(\frac{|r_1|}{n}+1\right)\sum_{m=0}^{n-1}\frac{\left|c_m^{(1)}\right|}{\varrho^{n-m}} \tag{7.65}$$

$$\le \frac{M(|r_1|+1)}{n}\sum_{m=0}^{n-1}\frac{\left|c_m^{(1)}\right|}{\varrho^{n-m}} \le \frac{K}{n}\sum_{m=0}^{n-1}\frac{\left|c_m^{(1)}\right|}{\varrho^{n-m}},$$

where

$$K = \max\{1,\ M(|r_1|+1)\}. \tag{7.66}$$

Since $K \ge 1$ we have $1 \le K \le K^2 \le \dots$, whence

$$\sum_{m=0}^{n-1} K^m \le nK^{n-1}. \tag{7.67}$$

Then, by induction on n,

$$\left|c_n^{(1)}\right| \le \left|c_0^{(1)}\right|\left(\frac{K}{\varrho}\right)^n \qquad (n = 0, 1, 2, \dots),$$

since, by (7.65) and (7.67),

$$\left|c_n^{(1)}\right| \le \frac{K}{n}\sum_{m=0}^{n-1}\frac{\left|c_m^{(1)}\right|}{\varrho^{n-m}} \le \frac{K}{n}\sum_{m=0}^{n-1}\frac{\left|c_0^{(1)}\right|}{\varrho^{n-m}}\left(\frac{K}{\varrho}\right)^m$$

$$= \frac{K\left|c_0^{(1)}\right|}{n\varrho^n}\sum_{m=0}^{n-1}K^m \le \frac{K\left|c_0^{(1)}\right|}{n\varrho^n}nK^{n-1} = \left|c_0^{(1)}\right|\left(\frac{K}{\varrho}\right)^n.$$

Hence $\left|c_n^{(1)}\right||z-z_0|^n \le \left|c_0^{(1)}\right|(K|z-z_0|/\varrho)^n$ $(n = 0, 1, 2, \dots)$. Since the geometric series

$$\sum_{n=0}^{\infty}\left(\frac{K|z-z_0|}{\varrho}\right)^n$$

converges for $|z-z_0| < \varrho/K$, the radius of convergence of (7.62) is $\ge \varrho/K$. Thus we have proved that the function

$$w_1(z) = (z-z_0)^{r_1}\sum_{n=0}^{\infty}c_n^{(1)}(z-z_0)^n \tag{7.68}$$

is a solution of (7.50) in the punctured disc

$$\mathcal{P}_1 = \{z \in \mathbb{C} \mid 0 < |z-z_0| < \varrho/K\}.$$

For any simply connected open set $\Omega \subset \mathcal{P}$ such that $\Omega \cap \mathcal{P}_1 \neq \varnothing$, by applying finitely many times Theorem 7.1 to the differential equation (7.50) on successive overlapping discs as we did in Sect. 7.2, we can analytically continue the solution $w_1(z)$ of (7.50) to the whole Ω. This shows that $w_1(z)$ can be analytically continued from $\mathcal{P}_1$ to $\mathcal{P}$. We conclude that the radius of convergence of (7.62) is $\geq R$, as claimed.

We now take $r = r_2$ in (7.55). In place of (7.60), we seek $c_n^{(2)}$ such that

$$f_0(r_2+n)c_n^{(2)} = -\sum_{m=0}^{n-1} f_{n-m}(r_2+m)c_m^{(2)} \qquad (n = 1, 2, 3, \ldots), \tag{7.69}$$

where

$$f_0(r_2+n) = n(n-s).$$

If $s = r_1 - r_2 = \nu$ for an integer $\nu \geq 1$ then $f_0(r_2+\nu) = \nu(\nu - s) = 0$, so that the recursion (7.69) does not determine $c_\nu^{(2)}, c_{\nu+1}^{(2)}, \ldots$. If $s = r_1 - r_2 = 0$ then (7.69) coincides with (7.60), and one finds again the series (7.62) previously obtained. Thus if $s \in \mathbb{N}$, through (7.55) we get only the solution (7.68) of (7.50), as expected.

If $s \notin \mathbb{N}$ then $f_0(r_2+n) = n(n-s) \neq 0$, and (7.69) yields $c_1^{(2)}, c_2^{(2)}, c_3^{(2)}, \ldots$ from an arbitrarily chosen $c_0^{(2)} \neq 0$. We now define

$$S = \sup_{n\geq 1} \frac{n}{|n-s|},$$

whence $S < +\infty$ since $n/|n-s| \to 1$ for $n \to \infty$. Then in place of (7.64) we get

$$|f_0(r_2+n)| = n\,|n-s| = \frac{n^2}{n/|n-s|} \geq \frac{n^2}{S}$$

and we can argue as in the previous case $r = r_1$, with (7.66) replaced by

$$K' = \max\{1, MS(|r_2|+1)\}.$$

Thus

$$|c_n^{(2)}| \leq |c_0^{(2)}| \left(\frac{K'}{\varrho}\right)^n \qquad (n = 0, 1, 2, \ldots),$$

and we conclude as above that the series

$$\sum_{n=0}^{\infty} c_n^{(2)} (z - z_0)^n$$

has radius of convergence $\geq R$. The solutions

$$w_1(z) = (z - z_0)^{r_1} \sum_{n=0}^{\infty} c_n^{(1)} (z - z_0)^n \quad \text{and} \quad w_2(z) = (z - z_0)^{r_2} \sum_{n=0}^{\infty} c_n^{(2)} (z - z_0)^n$$

of the differential equation (7.50) thus obtained are linearly independent by the argument we used to prove the linear independence of (7.37), or by remarking that the wronskian

$$\begin{vmatrix} w_1(z) & w_2(z) \\ w_1'(z) & w_2'(z) \end{vmatrix} = (r_2 - r_1) c_0^{(1)} c_0^{(2)} (z - z_0)^{r_1 + r_2 - 1} + \dots$$

cannot be identically zero.

It remains to find the second solution $w_2(z)$ of (7.50) when

$$s = r_1 - r_2 \in \mathbb{N}.$$

In this case $\lambda_1 = e^{2\pi i r_1} = \lambda_2 = e^{2\pi i r_2}$, and in accordance with (7.43) we seek the quotient $w_2(z)/w_1(z)$, where $w_1(z)$ is given by (7.68).

From (7.45), (7.48) and (7.49) we get

$$\frac{\mathrm{d}}{\mathrm{d}z} \log \left(w_1^2 \frac{\mathrm{d}}{\mathrm{d}z} \left(\frac{w_2}{w_1} \right) \right) = - \frac{A(z)}{z - z_0} = - \frac{a_0}{z - z_0} - \sum_{n=1}^{\infty} a_n (z - z_0)^{n-1}.$$

Integrating and taking exponentials we have

$$w_1^2 \frac{\mathrm{d}}{\mathrm{d}z} \left(\frac{w_2}{w_1} \right) = C (z - z_0)^{-a_0} \exp \left(- \sum_{n=1}^{\infty} \frac{a_n}{n} (z - z_0)^n \right),$$

where $C \neq 0$ is a constant. Thus, by (7.68),

$$\frac{\mathrm{d}}{\mathrm{d}z} \left(\frac{w_2}{w_1} \right) = C (z - z_0)^{-a_0 - 2r_1} \left(\sum_{n=0}^{\infty} c_n^{(1)} (z - z_0)^n \right)^{-2} \times \exp \left(- \sum_{n=1}^{\infty} \frac{a_n}{n} (z - z_0)^n \right). \quad (7.70)$$

Since $c_0^{(1)} \neq 0$, the function

$$\left(\sum_{n=0}^{\infty} c_n^{(1)} (z - z_0)^n \right)^{-2} \exp \left(- \sum_{n=1}^{\infty} \frac{a_n}{n} (z - z_0)^n \right)$$

is regular at z_0, and therefore in a disc $|z - z_0| < \varrho' \leq R$ has a Taylor expansion

$$\left(\sum_{n=0}^{\infty} c_n^{(1)}(z-z_0)^n\right)^{-2} \exp\left(-\sum_{n=1}^{\infty} \frac{a_n}{n}(z-z_0)^n\right) = \sum_{n=0}^{\infty} \alpha_n (z-z_0)^n,$$

with

$$\alpha_0 = \left(c_0^{(1)}\right)^{-2} \neq 0, \tag{7.71}$$

where the coefficients α_n are determined by the a_n in (7.49) and the $c_n^{(1)}$ given by the recurrence formula (7.60). Moreover, by (7.57), $r_1 + r_2 = 1 - a_0$, whence, by (7.59),

$$-a_0 - 2r_1 = r_1 + r_2 - 1 - 2r_1 = -(s+1).$$

Since in the present case s is an integer ≥ 0, (7.70) becomes, for $0 < |z - z_0| < \varrho'$,

$$\begin{aligned}\frac{\mathrm{d}}{\mathrm{d}z}\left(\frac{w_2}{w_1}\right) &= C(z-z_0)^{-(s+1)} \sum_{n=0}^{\infty} \alpha_n (z-z_0)^n \\ &= C\left(\frac{\alpha_0}{(z-z_0)^{s+1}} + \frac{\alpha_1}{(z-z_0)^s} + \ldots + \frac{\alpha_s}{z-z_0} + \alpha_{s+1} + \alpha_{s+2}(z-z_0) + \ldots\right).\end{aligned}$$

Integrating again we get, for constants $C \neq 0$ and C',

$$\begin{aligned}\frac{w_2}{w_1} = C\Bigg(&-\frac{\alpha_0}{s(z-z_0)^s} - \ldots - \frac{\alpha_{s-1}}{z-z_0} + \alpha_s \log(z-z_0) \\ &+ \alpha_{s+1}(z-z_0) + \frac{\alpha_{s+2}}{2}(z-z_0)^2 + \ldots + C'\Bigg) \\ &= H \log(z-z_0) + \psi_2(z),\end{aligned}$$

with $H = C\alpha_s$, where $\psi_2(z)$ is given in $0 < |z - z_0| < \varrho'$ by the Laurent expansion

$$\begin{aligned}\psi_2(z) = C\Bigg(&-\frac{\alpha_0}{s(z-z_0)^s} - \ldots - \frac{\alpha_{s-1}}{z-z_0} + C' \\ &+ \alpha_{s+1}(z-z_0) + \frac{\alpha_{s+2}}{2}(z-z_0)^2 + \ldots\Bigg).\end{aligned} \tag{7.72}$$

Therefore, as predicted in (7.43),

$$w_2(z) = w_1(z)\big(H \log(z-z_0) + \psi_2(z)\big), \tag{7.73}$$

where, by (7.71) and (7.72), $\psi_2(z)$ has a pole of order s at z_0 if $s > 0$, or is regular at z_0 if $s = 0$. This completes the proof of Fuchs' theorem. □

Chapter 8
Hypergeometric Functions

8.1 Totally Fuchsian Differential Equations

Let the functions $p_1(z)$ and $p_2(z)$ be one-valued and regular for any sufficiently large $|z|$, say for $|z| > R > 0$. Then the differential equation

$$w'' + p_1(z)\, w' + p_2(z)\, w = 0 \tag{8.1}$$

is said to satisfy Fuchs' conditions at the point $z = \infty$, whenever $Z = 0$ is a regular singular point of the differential equation obtained from (8.1) with the substitution $z = 1/Z$. Since

$$\begin{aligned} \frac{\mathrm{d}w}{\mathrm{d}z} &= -Z^2\,\frac{\mathrm{d}w}{\mathrm{d}Z}, \\ \frac{\mathrm{d}^2 w}{\mathrm{d}z^2} &= -Z^2\,\frac{\mathrm{d}}{\mathrm{d}Z}\Big(-Z^2\,\frac{\mathrm{d}w}{\mathrm{d}Z}\Big) = Z^4\,\frac{\mathrm{d}^2 w}{\mathrm{d}Z^2} + 2Z^3\,\frac{\mathrm{d}w}{\mathrm{d}Z}, \end{aligned} \tag{8.2}$$

the substitution $z = 1/Z$ changes (8.1) to

$$\frac{\mathrm{d}^2 w}{\mathrm{d}Z^2} + \left(\frac{2}{Z} - \frac{p_1(1/Z)}{Z^2}\right)\frac{\mathrm{d}w}{\mathrm{d}Z} + \frac{p_2(1/Z)}{Z^4}\, w = 0.$$

Hence Fuchs' conditions for (8.1) hold at $z = \infty$ if and only if $p_1(1/Z)$ has at least a simple zero and $p_2(1/Z)$ at least a double zero at $Z = 0$, i.e., if and only if $p_1(z)$ has at least a simple zero and $p_2(z)$ at least a double zero at $z = \infty$.

Definition 8.1 The differential equation (8.1) is said to be totally fuchsian when $p_1(z)$ and $p_2(z)$ are one-valued functions meromorphic in $\mathbb{C}$ and satisfying Fuchs' conditions at all the poles in $\mathbb{C}$ and at $z = \infty$.

C. Viola, *An Introduction to Special Functions*,
UNITEXT - La Matematica per il 3+2 102, DOI 10.1007/978-3-319-41345-7_8

If (8.1) is totally fuchsian, $p_1(z)$ and $p_2(z)$ vanish at $z = \infty$ and hence are regular in $\mathbb{C}\setminus\{z_1, \ldots, z_N\}$, where z_n $(n = 1, \ldots, N)$ is a pole for at least one of $p_1(z)$ and $p_2(z)$. Thus $p_1(z)$ and $p_2(z)$ can be decomposed as in (2.1), where the entire function $G(z)$ vanishes at $z = \infty$ and therefore is identically zero. Thus Fuchs' conditions at $z_1, \ldots, z_N$ yield

$$p_1(z) = \sum_{n=1}^{N} \frac{h_n}{z - z_n} \quad \text{and} \quad p_2(z) = \sum_{n=1}^{N} \frac{k_n z + l_n}{(z - z_n)^2}$$

with constants $h_n, k_n, l_n \in \mathbb{C}$, where $\sum_{n=1}^{N} k_n = 0$ since $p_2(z)$ vanishes at $z = \infty$ with order ≥ 2. We conclude that (8.1) is totally fuchsian if and only if $p_1(z)$ and $p_2(z)$ are rational functions of the form

$$p_1(z) = \frac{P_{N-1}(z)}{(z - z_1) \cdots (z - z_N)} \quad \text{and} \quad p_2(z) = \frac{Q_{2N-2}(z)}{(z - z_1)^2 \cdots (z - z_N)^2},$$

where the numerators are any polynomials satisfying $\deg P_{N-1} \leq N - 1$ and $\deg Q_{2N-2} \leq 2N - 2$.

8.2 The Hypergeometric Differential Equation

The prototype of totally fuchsian equation is the *hypergeometric differential equation*:

$$z(1 - z)w'' + (\gamma - (\alpha + \beta + 1)z)w' - \alpha\beta w = 0, \tag{8.3}$$

where α, β, γ are any complex parameters. The hypergeometric equation (8.3) was introduced by Euler, and later studied by Gauss, Kummer, Riemann and many other authors. The coefficients of (8.3) are

$$p_1(z) = \frac{\gamma - (\alpha + \beta + 1)z}{z(1 - z)}, \quad p_2(z) = -\frac{\alpha\beta}{z(1 - z)},$$

and Fuchs' conditions are plainly satisfied at $z = 0, 1, \infty$.

Using Frobenius' method described in the proof of Theorem 7.3, we seek solutions of (8.3) of the form (7.42) in the punctured disc $0 < |z| < 1$. With the notation (7.48), for $z_0 = 0$ we have

$$\begin{aligned} A(z) = \frac{\gamma - (\alpha + \beta + 1)z}{1 - z} &= (\gamma - (\alpha + \beta + 1)z)(1 + z + z^2 + \cdots) \\ &= \gamma + (\gamma - \alpha - \beta - 1)(z + z^2 + z^3 + \cdots), \end{aligned}$$

and

$$B(z) = -\frac{\alpha\beta z}{1-z} = -\alpha\beta(z + z^2 + z^3 + \cdots).$$

Thus at $z_0 = 0$ the auxiliary functions (7.54) are

$$\begin{aligned} f_0(t) &= t(t+\gamma-1), \\ f_\nu(t) &= (\gamma-\alpha-\beta-1)t - \alpha\beta \quad (\nu = 1, 2, 3, \ldots), \end{aligned} \tag{8.4}$$

and the indicial equation (7.57) is

$$r(r+\gamma-1) = 0. \tag{8.5}$$

According to (7.58) and the subsequent notation, we take

$$r_1 = 0 \text{ and } r_2 = 1-\gamma, \quad \text{if } \operatorname{Re}\gamma \geq 1,$$

or

$$r_1 = 1-\gamma \text{ and } r_2 = 0, \quad \text{if } \operatorname{Re}\gamma < 1.$$

If $s = r_1 - r_2 = \pm(\gamma - 1) \notin \mathbb{Z}$, i.e., if $\gamma \notin \mathbb{Z}$, there are two linearly independent solutions of (8.3) of the form (7.42), which we can write as

$$\sum_{n=0}^{\infty} c_n z^n \text{ with } c_0 = 1 \tag{8.6}$$

and

$$z^{1-\gamma} \sum_{n=0}^{\infty} \widetilde{c}_n z^n \text{ with } \widetilde{c}_0 = 1 \tag{8.7}$$

for $0 < |z| < 1$. In the case $\gamma \in \mathbb{Z}$, we have the solution (8.6) if $\gamma = 1, 2, 3, \ldots$, or the solution (8.7) if $\gamma = 0, -1, -2, \ldots$, while the second solution of (8.3), linearly independent of (8.6) or (8.7), in general contains a logarithmic term in accordance with (7.43).

Assuming $\gamma \in \mathbb{C}\backslash\{0, -1, -2, \ldots\}$, we compute the Taylor coefficients c_n of the solution (8.6). By (8.4), the recurrence formula (7.55) with $r = 0$ is

$$n(n+\gamma-1)c_n = -\sum_{m=0}^{n-1} \big((\gamma-\alpha-\beta-1)m - \alpha\beta\big)c_m. \tag{8.8}$$

From our choice $c_0 = 1$ we get $\gamma c_1 = \alpha\beta$, whence

$$c_1 = \frac{\alpha\beta}{\gamma},$$

and, for $n \geq 2$,

$$n(n+\gamma-1)c_n = \alpha\beta + \sum_{m=1}^{n-1} \big((\alpha+\beta-\gamma+1)m + \alpha\beta\big)c_m \tag{8.9}$$
$$= \alpha\beta + \sum_{m=1}^{n-1} \big((\alpha+m)(\beta+m) - m(m+\gamma-1)\big)c_m.$$

This yields

$$c_n = \frac{(\alpha)_n\,(\beta)_n}{(\gamma)_n\,n!} \quad (n = 0, 1, 2, \ldots), \tag{8.10}$$

where the Pochhammer symbols are defined by

$$\begin{aligned}(\alpha)_0 &= 1,\\ (\alpha)_n &= \alpha(\alpha+1)\cdots(\alpha+n-1) \quad (n = 1, 2, 3, \ldots),\end{aligned} \tag{8.11}$$

and similarly for $(\beta)_n$ and $(\gamma)_n$. We have seen that (8.10) holds for $n = 0, 1$. Assuming (8.10) for $n = 1, \ldots, q$, we get from (8.9)

$$(q+1)(q+\gamma)c_{q+1} = \alpha\beta + \sum_{m=1}^{q} \big((\alpha+m)(\beta+m) - m(m+\gamma-1)\big)\frac{(\alpha)_m\,(\beta)_m}{(\gamma)_m\,m!}$$
$$= \alpha\beta + \sum_{m=1}^{q} \frac{(\alpha)_{m+1}\,(\beta)_{m+1}}{(\gamma)_m\,m!} - \sum_{m=1}^{q} \frac{(\alpha)_m\,(\beta)_m}{(\gamma)_{m-1}\,(m-1)!}$$
$$= \frac{(\alpha)_{q+1}\,(\beta)_{q+1}}{(\gamma)_q\,q!},$$

whence

$$c_{q+1} = \frac{(\alpha)_{q+1}\,(\beta)_{q+1}}{(\gamma)_{q+1}\,(q+1)!},$$

which proves (8.10) by induction on n. Thus (8.6) is the Euler–Gauss *hypergeometric function* ${}_2F_1(\alpha, \beta; \gamma; z)$, defined by

$${}_2F_1(\alpha, \beta; \gamma; z) = \sum_{n=0}^{\infty} \frac{(\alpha)_n\,(\beta)_n}{(\gamma)_n}\,\frac{z^n}{n!} \quad (\gamma \neq 0, -1, -2, \ldots), \tag{8.12}$$

where the subscripts 2 and 1 of F indicate that the coefficients in the Taylor series (8.12) have two Pochhammer symbols in the numerator and one in the denominator.

If $\alpha = -N$ or $\beta = -N$, where $N \in \mathbb{N}$, the coefficients in the series (8.12) vanish for $n > N$, so that (8.12) is a polynomial in z of degree N. If $\alpha, \beta \neq 0, -1, -2, \ldots$ we get

$$\lim_{n\to\infty}\left|\frac{(\alpha)_n\,(\beta)_n}{(\gamma)_n\,n!}\cdot\frac{(\gamma)_{n+1}\,(n+1)!}{(\alpha)_{n+1}\,(\beta)_{n+1}}\right| = \lim_{n\to\infty}\left|\frac{(\gamma+n)(n+1)}{(\alpha+n)(\beta+n)}\right| = 1,$$

and the Taylor series (8.12) has radius of convergence 1.

We point out that several elementary functions are special cases of ${}_2F_1$. For example,

$$\begin{aligned}(1+z)^\alpha &= {}_2F_1(-\alpha,\beta;\beta;-z),\\ \frac{\log(1+z)}{z} &= {}_2F_1(1,1;2;-z),\\ \frac{\arctan z}{z} &= {}_2F_1(1/2,1;3/2;-z^2).\end{aligned}$$

We also remark that, by (8.12),

$$\begin{aligned}\frac{\mathrm{d}}{\mathrm{d}z}\,{}_2F_1(\alpha,\beta;\gamma;z) &= \sum_{n=1}^{\infty}\frac{(\alpha)_n\,(\beta)_n}{(\gamma)_n}\,\frac{z^{n-1}}{(n-1)!} = \sum_{n=0}^{\infty}\frac{(\alpha)_{n+1}\,(\beta)_{n+1}}{(\gamma)_{n+1}}\,\frac{z^n}{n!}\\ &= \frac{\alpha\beta}{\gamma}\sum_{n=0}^{\infty}\frac{(\alpha+1)_n\,(\beta+1)_n}{(\gamma+1)_n}\,\frac{z^n}{n!}\\ &= \frac{\alpha\beta}{\gamma}\,{}_2F_1(\alpha+1,\ \beta+1;\ \gamma+1;\ z),\end{aligned}$$

whence, by induction on m, we get the differentiation formula

$$\begin{aligned}&\frac{\mathrm{d}^m}{\mathrm{d}z^m}\,{}_2F_1(\alpha,\beta;\gamma;z)\\ &\quad= \frac{(\alpha)_m\,(\beta)_m}{(\gamma)_m}\,{}_2F_1(\alpha+m,\ \beta+m;\ \gamma+m;\ z)\quad (m=1,2,3,\dots).\end{aligned}\tag{8.13}$$

If $\gamma\in\mathbb{C}\setminus\{1,2,3,\dots\}$, we compute the coefficients $\widetilde{c}_n$ in (8.7). By (8.4) and the recurrence formula (7.55) with $r=1-\gamma$, we have

$$n(n-\gamma+1)\widetilde{c}_n = -\sum_{m=0}^{n-1}\big((\gamma-\alpha-\beta-1)(m-\gamma+1)-\alpha\beta\big)\widetilde{c}_m.\tag{8.14}$$

If in (8.14) we make the involutory substitution

$$\begin{cases}\alpha=\widetilde{\alpha}-\widetilde{\gamma}+1\\ \beta=\widetilde{\beta}-\widetilde{\gamma}+1\\ \gamma=2-\widetilde{\gamma},\end{cases}$$

we get (8.8) with α, β, γ, c replaced by $\widetilde{\alpha}, \widetilde{\beta}, \widetilde{\gamma}, \widetilde{c}$, respectively. Hence, by (8.10),

$$\widetilde{c}_n = \frac{(\widetilde{\alpha})_n\, (\widetilde{\beta})_n}{(\widetilde{\gamma})_n\, n!} = \frac{(\alpha-\gamma+1)_n\, (\beta-\gamma+1)_n}{(2-\gamma)_n\, n!} \qquad (n = 0, 1, 2, \ldots).$$

Thus the solution (8.7) is

$$z^{1-\gamma}\, {}_2F_1(\alpha-\gamma+1,\, \beta-\gamma+1;\, 2-\gamma;\, z) \qquad (\gamma \neq 1, 2, 3, \ldots). \tag{8.15}$$

8.3 Euler's Integral Representation of ${}_2F_1$

As we proved in Sect. 8.2, for $\alpha, \beta, \gamma \neq 0, -1, -2, \ldots$ the hypergeometric series (8.12) has radius of convergence 1. However, for any arbitrarily large $R > 1$ and for any simply connected open set Ω such that

$$\Omega \subset \{z \in \mathbb{C} \mid |z| < R,\ z \neq 1\}, \quad \Omega \cap \{z \in \mathbb{C} \mid |z| < 1\} \neq \varnothing,$$

if we apply finitely many times Theorem 7.1 to the hypergeometric differential equation (8.3) on successive overlapping discs as we did in Sect. 7.2, we can analytically continue the hypergeometric function ${}_2F_1(\alpha, \beta; \gamma; z)$ to the whole Ω. In particular, the hypergeometric function can be analytically continued to the cut plane $\mathbb{C}\backslash[1, +\infty)$, and this analytic continuation is again denoted by ${}_2F_1(\alpha, \beta; \gamma; z)$.

If $\operatorname{Re}\gamma > \operatorname{Re}\alpha > 0$, the analytic continuation of ${}_2F_1(\alpha, \beta; \gamma; z)$ to $\mathbb{C}\backslash[1, +\infty)$ can also be obtained through one of the most important formulae concerning ${}_2F_1$, namely the Euler integral representation, stated in the following

Theorem 8.1 (Euler) *If* $\operatorname{Re}\gamma > \operatorname{Re}\alpha > 0$ *and* $z \in \mathbb{C}\backslash[1, +\infty)$, *then*

$$\begin{aligned} {}_2F_1(\alpha, \beta; \gamma; z) &= \frac{\Gamma(\gamma)}{\Gamma(\alpha)\,\Gamma(\gamma-\alpha)} \int_0^1 \frac{t^{\alpha-1}(1-t)^{\gamma-\alpha-1}}{(1-tz)^{\beta}}\, \mathrm{d}t \\ &= \frac{1}{B(\alpha,\, \gamma-\alpha)} \int_0^1 \frac{t^{\alpha-1}(1-t)^{\gamma-\alpha-1}}{(1-tz)^{\beta}}\, \mathrm{d}t, \end{aligned} \tag{8.16}$$

where B and Γ are the Euler beta- and gamma-functions.

Proof The second equality in (8.16) is a consequence of (6.13). By (6.7),

$$\Gamma(\alpha+n) = \alpha(\alpha+1)\cdots(\alpha+n-1)\,\Gamma(\alpha) = (\alpha)_n\, \Gamma(\alpha), \tag{8.17}$$

and by (6.10) and (6.13),

$$\frac{\Gamma(\alpha+n)}{\Gamma(\gamma+n)} = \frac{B(\alpha+n,\gamma-\alpha)}{\Gamma(\gamma-\alpha)} = \frac{1}{\Gamma(\gamma-\alpha)}\int_0^1 t^{\alpha-1+n}(1-t)^{\gamma-\alpha-1}\,\mathrm{d}t.$$

Hence (8.17) yields

$$\frac{(\alpha)_n}{(\gamma)_n} = \frac{\Gamma(\alpha+n)}{\Gamma(\alpha)}\,\frac{\Gamma(\gamma)}{\Gamma(\gamma+n)} \tag{8.18}$$

$$= \frac{\Gamma(\gamma)}{\Gamma(\alpha)\,\Gamma(\gamma-\alpha)}\int_0^1 t^{\alpha-1+n}(1-t)^{\gamma-\alpha-1}\,\mathrm{d}t \qquad (n=0,1,2,\ldots).$$

Substituting this into (8.12) we get, for $|z|<1$,

$${}_2F_1(\alpha,\beta;\gamma;z) = \frac{\Gamma(\gamma)}{\Gamma(\alpha)\,\Gamma(\gamma-\alpha)}\sum_{n=0}^{\infty}(\beta)_n\frac{z^n}{n!}\int_0^1 t^{\alpha-1+n}(1-t)^{\gamma-\alpha-1}\,\mathrm{d}t \tag{8.19}$$

$$= \frac{\Gamma(\gamma)}{\Gamma(\alpha)\,\Gamma(\gamma-\alpha)}\int_0^1 t^{\alpha-1}(1-t)^{\gamma-\alpha-1}\sum_{n=0}^{\infty}\frac{(\beta)_n}{n!}(tz)^n\,\mathrm{d}t,$$

where the interchange of sum and integral is justified by absolute convergence, because

$$\sum_{n=0}^{\infty}|(\beta)_n|\frac{|z|^n}{n!}\int_0^1\left|t^{\alpha-1+n}(1-t)^{\gamma-\alpha-1}\right|\mathrm{d}t$$

$$\le \sum_{n=0}^{\infty}(|\beta|)_n\frac{|z|^n}{n!}\int_0^1 t^{\mathrm{Re}\,\alpha-1+n}(1-t)^{\mathrm{Re}\,\gamma-\mathrm{Re}\,\alpha-1}\,\mathrm{d}t$$

$$= \frac{\Gamma(\mathrm{Re}\,\alpha)\,\Gamma(\mathrm{Re}\,\gamma-\mathrm{Re}\,\alpha)}{\Gamma(\mathrm{Re}\,\gamma)}\,{}_2F_1(\mathrm{Re}\,\alpha,\,|\beta|;\,\mathrm{Re}\,\gamma;\,|z|).$$

For $0\le t\le 1$, $|z|<1$, the binomial series expansion yields

$$(1-tz)^{-\beta} = \sum_{n=0}^{\infty}(-1)^n\binom{-\beta}{n}(tz)^n$$

$$= \sum_{n=0}^{\infty}(-1)^n\frac{(-\beta)(-\beta-1)\cdots(-\beta-n+1)}{n!}(tz)^n = \sum_{n=0}^{\infty}\frac{(\beta)_n}{n!}(tz)^n.$$

Therefore, by (8.19),

$$ {}_2F_1(\alpha, \beta; \gamma; z) = \frac{\Gamma(\gamma)}{\Gamma(\alpha)\,\Gamma(\gamma-\alpha)} \int_0^1 \frac{t^{\alpha-1}(1-t)^{\gamma-\alpha-1}}{(1-tz)^{\beta}}\, dt. $$

If z varies in a neighbourhood of a point z_0 such that $|\arg(1-z_0)| < \pi$, i.e., $z_0 \in \mathbb{C}\setminus[1, +\infty)$, and t varies in $[0, 1]$, then

$$ \left|(1-tz)^{\beta}\right| = |1-tz|^{\operatorname{Re}\beta} \exp\big(-(\operatorname{Im}\beta)\arg(1-tz)\big) $$

is bounded from below by a positive constant. Thus, for $\operatorname{Re}\gamma > \operatorname{Re}\alpha > 0$, the integral in (8.16) is a regular function of z in $\mathbb{C}\setminus[1, +\infty)$. Hence (8.16) yields the analytic continuation of ${}_2F_1(\alpha, \beta; \gamma; z)$ from $|z| < 1$ to $z \in \mathbb{C}\setminus[1, +\infty)$. □

By (8.12),

$$ {}_2F_1(\alpha, \beta; \gamma; z) = {}_2F_1(\beta, \alpha; \gamma; z). \tag{8.20} $$

Therefore, if $\operatorname{Re}\gamma > \operatorname{Re}\beta > 0$ and $z \in \mathbb{C}\setminus[1, +\infty)$, (8.16) yields

$$ {}_2F_1(\alpha, \beta; \gamma; z) = \frac{\Gamma(\gamma)}{\Gamma(\beta)\,\Gamma(\gamma-\beta)} \int_0^1 \frac{t^{\beta-1}(1-t)^{\gamma-\beta-1}}{(1-tz)^{\alpha}}\, dt. \tag{8.21} $$

Moreover, if

$$ \operatorname{Re}\gamma > \max\{\operatorname{Re}\alpha, \operatorname{Re}\beta\} \text{ and } \min\{\operatorname{Re}\alpha, \operatorname{Re}\beta\} > 0, \tag{8.22} $$

then, by (8.16) and (8.21),

$$ \begin{aligned} {}_2F_1(\alpha, \beta; \gamma; z) &= \frac{\Gamma(\gamma)}{\Gamma(\alpha)\,\Gamma(\gamma-\alpha)} \int_0^1 \frac{t^{\alpha-1}(1-t)^{\gamma-\alpha-1}}{(1-tz)^{\beta}}\, dt \\ &= \frac{\Gamma(\gamma)}{\Gamma(\beta)\,\Gamma(\gamma-\beta)} \int_0^1 \frac{t^{\beta-1}(1-t)^{\gamma-\beta-1}}{(1-tz)^{\alpha}}\, dt, \end{aligned} $$

whence the integral transformation formula

$$ \int_0^1 \frac{t^{\alpha-1}(1-t)^{\gamma-\alpha-1}}{(1-tz)^{\beta}}\, dt = \frac{\Gamma(\alpha)\,\Gamma(\gamma-\alpha)}{\Gamma(\beta)\,\Gamma(\gamma-\beta)} \int_0^1 \frac{t^{\beta-1}(1-t)^{\gamma-\beta-1}}{(1-tz)^{\alpha}}\, dt \tag{8.23} $$

for any $z \in \mathbb{C}\setminus[1, +\infty)$.

Under the assumption (8.22), we give another proof of (8.23) independent of the hypergeometric function ${}_2F_1$. By (6.10) and (6.13) we have

$$\frac{\Gamma(\beta)\,\Gamma(\gamma-\beta)}{\Gamma(\gamma)}\int_0^1 \frac{t^{\alpha-1}(1-t)^{\gamma-\alpha-1}}{(1-tz)^{\beta}}\,dt \tag{8.24}$$

$$= B(\beta,\gamma-\beta)\int_0^1 \frac{t^{\alpha-1}(1-t)^{\gamma-\alpha-1}}{(1-tz)^{\beta}}\,dt$$

$$= \int_0^1\!\!\int_0^1 \frac{t^{\alpha-1}(1-t)^{\gamma-\alpha-1}u^{\beta-1}(1-u)^{\gamma-\beta-1}}{(1-tz)^{\beta}}\,dt\,du$$

$$= \int_0^1\!\!\int_0^1 \left(\frac{t}{1-t}\right)^{\alpha}\left(\frac{u}{(1-u)(1-tz)}\right)^{\beta}\big((1-t)(1-u)\big)^{\gamma}\,\frac{dt\,du}{t(1-t)u(1-u)}.$$

The last double integral suggests the change of variables

$$\frac{t}{1-t}=T,\qquad \frac{u}{(1-u)(1-tz)}=U,$$

which easily yields

$$(1-t)(1-u)=\frac{1}{1+T+U+TU(1-z)},\qquad \frac{dt\,du}{t(1-t)u(1-u)}=\frac{dT\,dU}{TU}.$$

Thus, if $z\in\mathbb{R}$, $z<1$, from (8.24) we get

$$\frac{\Gamma(\beta)\,\Gamma(\gamma-\beta)}{\Gamma(\gamma)}\int_0^1 \frac{t^{\alpha-1}(1-t)^{\gamma-\alpha-1}}{(1-tz)^{\beta}}\,dt$$

$$= \int_0^{+\infty}\!\!\int_0^{+\infty} \frac{T^{\alpha-1}\,U^{\beta-1}}{\big(1+T+U+TU(1-z)\big)^{\gamma}}\,dT\,dU,$$

which is symmetric in α and β. This proves (8.23) for $z\in\mathbb{R}$, $z<1$, and hence for any $z\in\mathbb{C}\backslash[1,+\infty)$ by analytic continuation.

If neither $\operatorname{Re}\gamma>\operatorname{Re}\alpha>0$ nor $\operatorname{Re}\gamma>\operatorname{Re}\beta>0$, the analytic continuation of ${}_2F_1(\alpha,\beta;\gamma;z)$ to $\mathbb{C}\backslash[1,+\infty)$ can be reduced to (8.16) or (8.21) as follows. A straightforward computation yields, for any $\alpha,\beta\in\mathbb{C}$, $\gamma\in\mathbb{C}\backslash\{0,-1,-2,\dots\}$, and $k\in\mathbb{N}$, $k\geq 1$,

$$\gamma(\gamma-\beta+1)\,\frac{(\alpha+1)_k\,(\beta)_k}{(\gamma+2)_k\,k!}+\beta\gamma\,\frac{(\alpha+1)_k\,(\beta+1)_k}{(\gamma+2)_k\,k!}$$
$$-\,\beta(\gamma-\alpha)\,\frac{(\alpha+1)_{k-1}\,(\beta+1)_{k-1}}{(\gamma+2)_{k-1}\,(k-1)!}\;=\;\gamma(\gamma+1)\frac{(\alpha)_k\,(\beta)_k}{(\gamma)_k\,k!}.$$

Multiplying by z^k with $|z|<1$ and summing for $k\geq 1$, we get

$$\begin{aligned}&\gamma(\gamma-\beta+1)\big({}_2F_1(\alpha+1,\,\beta;\,\gamma+2;\,z)-1\big)+\\&\beta\gamma\big({}_2F_1(\alpha+1,\,\beta+1;\,\gamma+2;\,z)-1\big)\\&-\beta(\gamma-\alpha)z\,{}_2F_1(\alpha+1,\,\beta+1;\,\gamma+2;\,z)\\&=\gamma(\gamma+1)\big({}_2F_1(\alpha,\beta;\gamma;z)-1\big),\end{aligned}$$

whence

$$\begin{aligned}{}_2F_1(\alpha,\beta;\gamma;z)=&\,\frac{\gamma-\beta+1}{\gamma+1}\,{}_2F_1(\alpha+1,\,\beta;\,\gamma+2;\,z)\\&+\frac{\beta(\gamma-(\gamma-\alpha)z)}{\gamma(\gamma+1)}\,{}_2F_1(\alpha+1,\,\beta+1;\,\gamma+2;\,z).\end{aligned}\tag{8.25}$$

By applying (8.25) successively n times, we get

$${}_2F_1(\alpha,\beta;\gamma;z)\;=\sum_{m=0}^{n}c_{mn}(\alpha,\beta;\gamma;z)\,{}_2F_1(\alpha+n,\,\beta+m;\,\gamma+2n;\,z),\tag{8.26}$$

where, as is easy to prove by induction on n, for every m,n with $0\leq m\leq n$

$$(\gamma)_{2n}\,c_{mn}(\alpha,\beta;\gamma;z)\ \text{ is a polynomial in }\alpha,\beta,\gamma,z.\tag{8.27}$$

If we choose

$$n\;>\;\max\{-\operatorname{Re}\alpha,\;\operatorname{Re}(\alpha-\gamma)\},\tag{8.28}$$

then $\operatorname{Re}(\gamma+2n)>\operatorname{Re}(\alpha+n)>0$, whence, by (8.16) and (8.26),

$${}_2F_1(\alpha,\beta;\gamma;z)=$$
$$\sum_{m=0}^{n}c_{mn}(\alpha,\beta;\gamma;z)\,\frac{\Gamma(\gamma+2n)}{\Gamma(\alpha+n)\,\Gamma(\gamma-\alpha+n)}\int_0^1\frac{t^{\alpha+n-1}(1-t)^{\gamma-\alpha+n-1}}{(1-tz)^{\beta+m}}\,\mathrm{d}t.$$

Here each term in the sum on the right-hand side is a regular function of z in $\mathbb{C}\backslash[1,+\infty)$.

Theorem 8.2 *Let $\gamma \neq 0, -1, -2, \ldots$, and let*

$$\operatorname{Re}(\gamma - \alpha - \beta) > 0. \tag{8.29}$$

Then the hypergeometric series (8.12) *is totally convergent in the closed disc $|z| \le 1$. Moreover*

$$\sum_{n=0}^{\infty} \frac{(\alpha)_n\,(\beta)_n}{(\gamma)_n\, n!} = \lim_{z\to 1-} {}_2F_1(\alpha, \beta; \gamma; z) = \frac{\Gamma(\gamma)\,\Gamma(\gamma-\alpha-\beta)}{\Gamma(\gamma-\alpha)\,\Gamma(\gamma-\beta)}. \tag{8.30}$$

Proof If $\alpha = -N$ or $\beta = -N$ with $N \in \mathbb{N}$, (8.12) is a polynomial, so the total convergence is trivial. Otherwise, for $|z| \le 1$ we have

$$\left|\frac{(\alpha)_n\,(\beta)_n}{(\gamma)_n}\,\frac{z^n}{n!}\right| \le \left|\frac{\Gamma(\gamma)}{\Gamma(\alpha)\,\Gamma(\beta)}\right| \cdot \left|\frac{\Gamma(\alpha+n)\,\Gamma(\beta+n)}{\Gamma(\gamma+n)\,\Gamma(n+1)}\right|, \tag{8.31}$$

since $n! = \Gamma(n+1)$ by (6.9) and $(\alpha)_n = \Gamma(\alpha+n)/\Gamma(\alpha)$ by (8.17), and similarly for $(\beta)_n$ and $(\gamma)_n$. By (6.40) with $s = 0$ and $z = n$ we get, for $n \to \infty$,

$$\log \Gamma(\alpha+n) = \left(n + \alpha - \frac{1}{2}\right)\log n - n + \log\sqrt{2\pi} + O\!\left(\frac{1}{n}\right).$$

Using the same asymptotic formula with α replaced by β, γ or 1 we obtain

$$\log \frac{\Gamma(\alpha+n)\,\Gamma(\beta+n)}{\Gamma(\gamma+n)\,\Gamma(n+1)} = (\alpha+\beta-\gamma-1)\log n + O\!\left(\frac{1}{n}\right).$$

Taking exponentials and then absolute values, we get by (6.3)

$$\left|\frac{\Gamma(\alpha+n)\,\Gamma(\beta+n)}{\Gamma(\gamma+n)\,\Gamma(n+1)}\right| = \left|n^{\alpha+\beta-\gamma-1}\right|\left(1 + O\!\left(\frac{1}{n}\right)\right) = n^{\operatorname{Re}(\alpha+\beta-\gamma)-1}\left(1 + O\!\left(\frac{1}{n}\right)\right).$$

By (8.29),

$$\sum_n n^{\operatorname{Re}(\alpha+\beta-\gamma)-1} < +\infty.$$

Hence, by (8.31), the series (8.12) is totally convergent for $|z| \le 1$. In particular, (8.12) is uniformly convergent for $0 \le z \le 1$, whence

$$\lim_{z\to 1-} {}_2F_1(\alpha, \beta; \gamma; z) = \sum_{n=0}^{\infty} \frac{(\alpha)_n\,(\beta)_n}{(\gamma)_n\, n!}. \tag{8.32}$$

If

$$\operatorname{Re}\gamma > \operatorname{Re}\alpha > 0, \tag{8.33}$$

under the assumption (8.29) we have, for $0 < t < 1,\ 0 \le z \le 1$,

$$\left|\frac{t^{\alpha-1}(1-t)^{\gamma-\alpha-1}}{(1-tz)^{\beta}}\right| \le t^{\operatorname{Re}\alpha-1}(1-t)^{\delta-1} \tag{8.34}$$

with

$$\delta = \begin{cases} \operatorname{Re}(\gamma-\alpha), & \text{if } \operatorname{Re}\beta \le 0 \\ \operatorname{Re}(\gamma-\alpha-\beta), & \text{if } \operatorname{Re}\beta > 0, \end{cases}$$

whence $\delta > 0$. Therefore

$$\int_0^1 t^{\operatorname{Re}\alpha-1}(1-t)^{\delta-1}\,\mathrm{d}t < +\infty. \tag{8.35}$$

By (8.34) and (8.35), the integral in (8.16) is uniformly convergent for $0 \le z \le 1$. Hence, making $z \to 1-$ in (8.16), we can interchange limit and integral. By (6.10) and (6.13) we get

$$\begin{aligned}\lim_{z\to 1-} {}_2F_1(\alpha,\beta;\gamma;z) &= \frac{\Gamma(\gamma)}{\Gamma(\alpha)\,\Gamma(\gamma-\alpha)}\int_0^1 t^{\alpha-1}(1-t)^{\gamma-\alpha-\beta-1}\,\mathrm{d}t \\ &= \frac{\Gamma(\gamma)}{\Gamma(\alpha)\,\Gamma(\gamma-\alpha)}B(\alpha,\gamma-\alpha-\beta) = \frac{\Gamma(\gamma)\,\Gamma(\gamma-\alpha-\beta)}{\Gamma(\gamma-\alpha)\,\Gamma(\gamma-\beta)}.\end{aligned} \tag{8.36}$$

If instead of (8.33) we assume the weaker inequalities

$$\operatorname{Re}\alpha > -1 \text{ and } \operatorname{Re}(\gamma-\alpha) > -1,$$

then $\operatorname{Re}(\alpha+1) > 0$, $\operatorname{Re}((\gamma+2)-(\alpha+1)) = \operatorname{Re}(\gamma-\alpha+1) > 0$, and, by (8.29), $\operatorname{Re}((\gamma+2)-(\alpha+1)-\beta) = \operatorname{Re}(\gamma-\alpha-\beta)+1 > 0$, so that the two hypergeometric functions on the right-hand side of (8.25) satisfy (8.29) and (8.33). Applying (8.36) to such hypergeometric functions, by (6.6) we obtain

$$\begin{aligned}\lim_{z\to 1-} {}_2F_1(\alpha,\beta;\gamma;z) &= \frac{\gamma-\beta+1}{\gamma+1}\,\frac{\Gamma(\gamma+2)\,\Gamma(\gamma-\alpha-\beta+1)}{\Gamma(\gamma-\alpha+1)\,\Gamma(\gamma-\beta+2)} \\ &\quad + \frac{\alpha\beta}{\gamma(\gamma+1)}\,\frac{\Gamma(\gamma+2)\,\Gamma(\gamma-\alpha-\beta)}{\Gamma(\gamma-\alpha+1)\,\Gamma(\gamma-\beta+1)} \\ &= \frac{\Gamma(\gamma)\,\Gamma(\gamma-\alpha-\beta)}{\Gamma(\gamma-\alpha)\,\Gamma(\gamma-\beta)}.\end{aligned}$$

Iterating this process n times, under the assumptions (8.29) and

$$\operatorname{Re}\alpha > -n \text{ and } \operatorname{Re}(\gamma-\alpha) > -n$$

we get

$$\lim_{z\to 1-} {}_2F_1(\alpha, \beta; \gamma; z) = \frac{\Gamma(\gamma)\,\Gamma(\gamma-\alpha-\beta)}{\Gamma(\gamma-\alpha)\,\Gamma(\gamma-\beta)}. \tag{8.37}$$

Since n is arbitrary, we can choose n satisfying (8.28). Hence (8.37) holds under the assumption (8.29) only.

From (8.32) and (8.37) we get (8.30). □

8.4 ${}_2F_1$ as a Function of the Parameters

Theorem 8.3 *For any fixed* $z \in \mathbb{C}\backslash[1, +\infty)$,

$$\frac{{}_2F_1(\alpha, \beta; \gamma; z)}{\Gamma(\gamma)}$$

is an entire function of α, β *and* γ.

Proof If

$$\operatorname{Re}\gamma > \operatorname{Re}\alpha > 0, \tag{8.38}$$

from (8.16) we get

$$\frac{{}_2F_1(\alpha, \beta; \gamma; z)}{\Gamma(\gamma)} = \frac{1}{\Gamma(\alpha)}\,\frac{1}{\Gamma(\gamma-\alpha)}\int_0^1 \frac{t^{\alpha-1}(1-t)^{\gamma-\alpha-1}}{(1-tz)^{\beta}}\,dt. \tag{8.39}$$

For any fixed $z \in \mathbb{C}\backslash[1, +\infty)$, the integral on the right-hand side of (8.39) is uniformly convergent for

$$\varepsilon \le \operatorname{Re}\alpha \le A,\ \ |\beta| \le B \ \text{ and } \ \varepsilon \le \operatorname{Re}(\gamma-\alpha) \le C,$$

where A, B and C can be taken arbitrarily large and $\varepsilon > 0$ arbitrarily small. Moreover, for any $0 < t < 1$ the integrand is plainly an entire function of α, β and γ, and $1/\Gamma(\alpha)$ and $1/\Gamma(\gamma-\alpha)$ are entire functions of α and γ by Theorem 6.3. Hence (8.39) is an entire function of β, and a regular function of α and γ for $\operatorname{Re}\alpha > 0$ and $\operatorname{Re}(\gamma-\alpha) > 0$, i.e., in the region (8.38). If (8.38) does not hold, by (8.26) and (8.17) we get

$$\frac{{}_2F_1(\alpha, \beta; \gamma; z)}{\Gamma(\gamma)} = \sum_{m=0}^{n} (\gamma)_{2n}\, c_{mn}(\alpha, \beta; \gamma; z)\, \frac{{}_2F_1(\alpha+n, \beta+m;\ \gamma+2n;\ z)}{\Gamma(\gamma+2n)}. \tag{8.40}$$

By the previous argument and by (8.27), each term in the sum on the right-hand side of (8.40) is an entire function of β, and a regular function of α and γ for

$$\operatorname{Re}\alpha > -n \ \text{ and } \ \operatorname{Re}(\gamma-\alpha) > -n.$$

Since n can be taken arbitrarily large, choosing n as in (8.28) we conclude that (8.40) is an entire function of α, β and γ. □

The six functions

$${}_2F_1(\alpha \pm 1, \beta; \gamma; z), \quad {}_2F_1(\alpha, \beta \pm 1; \gamma; z), \quad {}_2F_1(\alpha, \beta; \gamma \pm 1; z)$$

are called *contiguous* to ${}_2F_1(\alpha, \beta; \gamma; z)$. The function ${}_2F_1(\alpha, \beta; \gamma; z)$ and any two functions contiguous to it are related by homogeneous three-terms linear identities, whose coefficients are polynomials in α, β, γ, z of partial degrees 0 or 1 in each of α, β, z, and of partial degrees 0, 1 or 2 in γ. Thus there are $\binom{6}{2} = 15$ such identities, due to Gauss, which can be proved by direct substitution of the hypergeometric series (8.12). The full list of the fifteen Gauss contiguity formulae can be found in vol. 1, pp. 103–104, of the treatise by A. Erdélyi et al. quoted in the bibliography.

To simplify the notation, it is customary to abbreviate

$$\begin{aligned} {}_2F_1(\alpha, \beta; \gamma; z) &= F, & {}_2F_1(\alpha \pm 1, \beta; \gamma; z) &= F(\alpha \pm 1), \\ {}_2F_1(\alpha, \beta \pm 1; \gamma; z) &= F(\beta \pm 1), & {}_2F_1(\alpha, \beta; \gamma \pm 1; z) &= F(\gamma \pm 1). \end{aligned}$$

As an example, we prove the contiguity formula relating F, $F(\alpha+1)$, $F(\beta+1)$, namely

$$(\alpha-\beta)F - \alpha\, F(\alpha+1) + \beta\, F(\beta+1) = 0. \tag{8.41}$$

Multiplying the identity

$$(\alpha-\beta) - (\alpha+n) + (\beta+n) = 0$$

by

$$\frac{(\alpha)_n\,(\beta)_n}{(\gamma)_n}\,\frac{z^n}{n!}$$

with $|z| < 1$, we get

$$\Big((\alpha-\beta)(\alpha)_n(\beta)_n - \alpha(\alpha+1)_n(\beta)_n + (\alpha)_n\beta(\beta+1)_n\Big)\frac{z^n}{(\gamma)_n\, n!} = 0.$$

Summing for $n \geq 0$ we obtain (8.41).

8.5 Linear Transformations

The group of fractional linear transformations

$$Z = \frac{az+b}{cz+d} \qquad (a, b, c, d \in \mathbb{C}, \ ad - bc \neq 0)$$

carrying the points $z = 0, 1, \infty$ onto the points $Z = 0, 1, \infty$ in any order is obviously isomorphic to the symmetric group $\mathfrak{S}_3$ of order $3! = 6$. Its elements are easily seen to be

$$\begin{aligned} &Z = z, \quad Z = 1 - z, \ Z = \frac{1}{z}, \\ &Z = \frac{z}{z-1}, \ Z = \frac{z-1}{z}, \ Z = \frac{1}{1-z}. \end{aligned} \tag{8.42}$$

If we apply any one of the transformations (8.42) to the differential equation (8.3), we change (8.3) to a differential equation with regular singular points at $Z = 0, 1, \infty$, which is again of type (8.3) where the parameters α, β, γ are suitably transformed.

We begin with the simplest case, i.e., with the transformation $Z = 1 - z$. Since

$$\frac{\mathrm{d}w}{\mathrm{d}z} = \frac{\mathrm{d}w}{\mathrm{d}Z}\frac{\mathrm{d}Z}{\mathrm{d}z} = -\frac{\mathrm{d}w}{\mathrm{d}Z}$$

and

$$\frac{\mathrm{d}^2 w}{\mathrm{d}z^2} = \frac{\mathrm{d}}{\mathrm{d}z}\Big(-\frac{\mathrm{d}w}{\mathrm{d}Z}\Big) = -\frac{\mathrm{d}^2 w}{\mathrm{d}Z^2}\frac{\mathrm{d}Z}{\mathrm{d}z} = \frac{\mathrm{d}^2 w}{\mathrm{d}Z^2},$$

(8.3) becomes

$$Z(1-Z)\frac{\mathrm{d}^2 w}{\mathrm{d}Z^2} + \big(\alpha+\beta-\gamma+1-(\alpha+\beta+1)Z\big)\frac{\mathrm{d}w}{\mathrm{d}Z} - \alpha\beta w = 0, \tag{8.43}$$

which is the hypergeometric differential equation with the change of parameters

$$(\alpha, \beta, \gamma) \longmapsto (\alpha, \beta, \alpha+\beta-\gamma+1)$$

in (8.3).

Applying the solutions (8.12) and (8.15) of (8.3) to the differential equation (8.43), and then substituting $Z = 1 - z$, we get two further solutions of (8.3) in the cut plane $\mathbb{C}\backslash(-\infty, 0\,]$, namely

$${}_2F_1(\alpha, \beta; \alpha+\beta-\gamma+1; 1-z) \qquad (\gamma-\alpha-\beta \neq 1, 2, 3, \dots) \tag{8.44}$$

and, combining with (8.20),

$$(1-z)^{\gamma-\alpha-\beta}\,{}_2F_1(\gamma-\alpha,\,\gamma-\beta;\;\gamma-\alpha-\beta+1;\;1-z) \qquad (\gamma-\alpha-\beta\neq 0,-1,-2,\dots), \tag{8.45}$$

which are linearly independent if $\gamma-\alpha-\beta\notin\mathbb{Z}$.

By (8.2), the transformation $Z=1/z$ changes (8.3) to

$$Z(1-Z)\frac{\mathrm{d}^2w}{\mathrm{d}Z^2}+(1-\alpha-\beta+(\gamma-2)Z)\frac{\mathrm{d}w}{\mathrm{d}Z}+\frac{\alpha\beta}{Z}\,w=0.$$

Here we apply the change of function $w=Z^{\alpha}W$. We easily obtain the equation

$$Z(1-Z)\frac{\mathrm{d}^2W}{\mathrm{d}Z^2}+(\alpha-\beta+1-(2\alpha-\gamma+2)Z)\frac{\mathrm{d}W}{\mathrm{d}Z}-\alpha(\alpha-\gamma+1)W=0, \tag{8.46}$$

which is of type (8.3) with the change of parameters

$$(\alpha,\beta,\gamma)\longmapsto(\alpha,\;\alpha-\gamma+1,\;\alpha-\beta+1).$$

Applying (8.12) and (8.15) to (8.46) and then substituting $Z=1/z,\ W=Z^{-\alpha}w=z^{\alpha}w$, we get two new solutions of (8.3) in the region $\mathbb{C}\backslash[\,0,1\,]$, namely

$$z^{-\alpha}\,{}_2F_1(\alpha,\;\alpha-\gamma+1;\;\alpha-\beta+1;\;1/z) \qquad (\beta-\alpha\neq 1,2,3,\dots) \tag{8.47}$$

and

$$z^{-\beta}\,{}_2F_1(\beta,\;\beta-\gamma+1;\;\beta-\alpha+1;\;1/z) \qquad (\beta-\alpha\neq 0,-1,-2,\dots), \tag{8.48}$$

which are linearly independent if $\beta-\alpha\notin\mathbb{Z}$.

The remaining three transformations in the second row of (8.42) yield no further solutions of (8.3). This is a consequence of the involutory transformation formula (8.49) below.

Theorem 8.4 (Euler) *For any* $\alpha,\beta\in\mathbb{C},\ \gamma\in\mathbb{C}\backslash\{0,-1,-2,\dots\}$ *and* $z\in\mathbb{C}\backslash[1,+\infty)$ *we have*

$${}_2F_1(\alpha,\beta;\gamma;z)\;=\;(1-z)^{-\alpha}\,{}_2F_1(\alpha,\;\gamma-\beta;\;\gamma;\;z/(z-1)) \tag{8.49}$$

and

$${}_2F_1(\alpha,\beta;\gamma;z)\;=\;(1-z)^{\gamma-\alpha-\beta}\,{}_2F_1(\gamma-\alpha,\;\gamma-\beta;\;\gamma;\;z). \tag{8.50}$$

Proof Let

$$B=\gamma-\beta,\qquad Z=\frac{z}{z-1}. \tag{8.51}$$

Clearly $\operatorname{Re}\gamma>\operatorname{Re}\beta>0$ if and only if $\operatorname{Re}\gamma>\operatorname{Re}B>0$. We temporarily assume this. For any $z\in\mathbb{C}\backslash[1,+\infty)$, we substitute $t=1-T$ in the integral (8.21). We get

$$
\begin{aligned}
{}_2F_1(\alpha, \beta; \gamma; z) &= \frac{\Gamma(\gamma)}{\Gamma(\beta)\,\Gamma(\gamma-\beta)} \int_0^1 \frac{T^{\gamma-\beta-1}(1-T)^{\beta-1}}{(1-z+Tz)^{\alpha}}\,\mathrm{d}T \\
&= (1-z)^{-\alpha} \frac{\Gamma(\gamma)}{\Gamma(B)\,\Gamma(\gamma-B)} \int_0^1 \frac{T^{B-1}(1-T)^{\gamma-B-1}}{(1-TZ)^{\alpha}}\,\mathrm{d}T \\
&= (1-z)^{-\alpha}\,{}_2F_1(\alpha, B; \gamma; Z),
\end{aligned}
$$

whence (8.49). Since, by Theorem 8.3, for any $z \in \mathbb{C}\backslash[1,+\infty)$ both sides of (8.49) divided by $\Gamma(\gamma)$ are entire functions of β and γ, by analytic continuation on β and γ the restrictive assumption $\operatorname{Re}\gamma > \operatorname{Re}\beta > 0$ can be dropped. Thus (8.49) holds for any $\alpha, \beta \in \mathbb{C}$, for any $\gamma \in \mathbb{C}\backslash\{0, -1, -2, \dots\}$ and for any $z \in \mathbb{C}\backslash[1, +\infty)$.

By the symmetry property (8.20) and by (8.49) we get

$$
{}_2F_1(\alpha, \beta; \gamma; z) = (1-z)^{-\beta}\,{}_2F_1(\gamma-\alpha, \beta; \gamma; z/(z-1)). \tag{8.52}
$$

With the notation (8.51) we have $Z \in \mathbb{C}\backslash[1,+\infty)$ if and only if $z \in \mathbb{C}\backslash[1,+\infty)$. If we apply first (8.49) and then (8.52) we obtain

$$
\begin{aligned}
{}_2F_1(\alpha, \beta; \gamma; z) &= (1-z)^{-\alpha}\,{}_2F_1(\alpha, B; \gamma; Z) \\
&= (1-z)^{-\alpha}(1-Z)^{-B}\,{}_2F_1(\gamma-\alpha, B; \gamma; Z/(Z-1)),
\end{aligned}
$$

whence (8.50). □

We remark that the transformation formula (8.50) is noteworthy because, unlike (8.49) or other identities, it relates the values of two hypergeometric functions at the same point z.

We now combine (8.49) with the action of the group (8.42). The transformation $Z = z/(z-1)$, together with the change of function $w = (1-Z)^{\alpha} W$, changes (8.3) to

$$
Z(1-Z)\frac{\mathrm{d}^2 W}{\mathrm{d}Z^2} + \big(\gamma - (\alpha - \beta + \gamma + 1)Z\big)\frac{\mathrm{d}W}{\mathrm{d}Z} - \alpha(\gamma-\beta)W = 0, \tag{8.53}
$$

corresponding to the change of parameters in (8.3)

$$
(\alpha, \beta, \gamma) \longmapsto (\alpha, \gamma-\beta, \gamma).
$$

We apply (8.12) and (8.15) to (8.53), and then substitute $Z = z/(z-1)$, $W = (1-Z)^{-\alpha}w = (1-z)^{\alpha}w$. Since, as we remarked, $Z \in \mathbb{C}\backslash[1,+\infty)$ if and only if $z \in \mathbb{C}\backslash[1,+\infty)$, we get the following solutions of (8.3) in the cut plane $\mathbb{C}\backslash[1,+\infty)$:

$$
(1-z)^{-\alpha}\,{}_2F_1(\alpha, \gamma-\beta; \gamma; z/(z-1)) \qquad (\gamma \neq 0, -1, -2, \dots) \tag{8.54}
$$

and

$$(-z)^{1-\gamma}(1-z)^{\gamma-\alpha-1}\,{}_2F_1(\alpha-\gamma+1,\,1-\beta;\,2-\gamma;\,z/(z-1)) \qquad (8.55)$$
$$(\gamma \neq 1, 2, 3, \dots).$$

By (8.49), the functions (8.12) and (8.54) coincide. Moreover (8.49) yields, for $\gamma \neq 2, 3, 4, \dots,$

$$\begin{aligned} &z^{1-\gamma}\,{}_2F_1(\alpha-\gamma+1,\,\beta-\gamma+1;\,2-\gamma;\,z) \\ &\qquad = z^{1-\gamma}(1-z)^{\gamma-\alpha-1}\,{}_2F_1(\alpha-\gamma+1,\,1-\beta;\,2-\gamma;\,z/(z-1)) \end{aligned} \qquad (8.56)$$

(note that for $\gamma = 1$ the identities (8.49) and (8.56) are the same). By (8.56), we see that (8.15) and (8.55) coincide up to the constant factor $(-1)^{1-\gamma}$. Hence the transformation $Z = z/(z-1)$ yields no new solutions of (8.3).

We remark that, by applying (8.49), (8.45) becomes

$$z^{\alpha-\gamma}(1-z)^{\gamma-\alpha-\beta}\,{}_2F_1(\gamma-\alpha,\,1-\alpha;\,\gamma-\alpha-\beta+1;\,(z-1)/z). \qquad (8.57)$$

The transformation $Z = (z-1)/z$, together with the change of function $w = (1-Z)^\alpha W$, changes (8.3) to

$$Z(1-Z)\frac{\mathrm{d}^2W}{\mathrm{d}Z^2} + (\alpha+\beta-\gamma+1-(2\alpha-\gamma+2)Z)\frac{\mathrm{d}W}{\mathrm{d}Z} - \alpha(\alpha-\gamma+1)W = 0,$$

corresponding to the change of parameters

$$(\alpha, \beta, \gamma) \longmapsto (\alpha,\, \alpha-\gamma+1,\, \alpha+\beta-\gamma+1).$$

This equation yields the following solutions of (8.3) in the cut plane $\mathbb{C}\backslash(-\infty, 0]$:

$$z^{-\alpha}\,{}_2F_1(\alpha,\,\alpha-\gamma+1;\,\alpha+\beta-\gamma+1;\,(z-1)/z) \quad (\gamma-\alpha-\beta \neq 1, 2, 3, \dots) \qquad (8.58)$$

and

$$z^{\beta-\gamma}(z-1)^{\gamma-\alpha-\beta}\,{}_2F_1(\gamma-\beta,\,1-\beta;\,\gamma-\alpha-\beta+1;\,(z-1)/z) \qquad (8.59)$$
$$(\gamma-\alpha-\beta \neq 0, -1, -2, \dots).$$

By (8.49), the solution (8.58) coincides with (8.44), and (8.59) coincides with (8.45) up to the constant factor $(-1)^{\gamma-\alpha-\beta}$. Moreover, since (8.45) equals (8.57), (8.59) can also be written as

$$z^{\alpha-\gamma}(z-1)^{\gamma-\alpha-\beta}\,{}_2F_1(\gamma-\alpha,\,1-\alpha;\,\gamma-\alpha-\beta+1;\,(z-1)/z) \qquad (8.60)$$
$$(\gamma-\alpha-\beta \neq 0, -1, -2, \dots).$$

Finally, the transformation $Z = 1/(1-z)$, together with the change of function $w = Z^{\alpha} W$, changes (8.3) to

$$Z(1-Z)\frac{\mathrm{d}^2 W}{\mathrm{d}Z^2} + (\alpha - \beta + 1 - (\alpha - \beta + \gamma + 1)Z)\frac{\mathrm{d}W}{\mathrm{d}Z} - \alpha(\gamma - \beta)W = 0,$$

corresponding to the change of parameters

$$(\alpha, \beta, \gamma) \longmapsto (\alpha,\ \gamma - \beta,\ \alpha - \beta + 1).$$

This equation yields the solutions of (8.3) in the region $\mathbb{C}\backslash[0,1]$:

$$(1-z)^{-\alpha}\,{}_2F_1(\alpha,\ \gamma-\beta;\ \alpha-\beta+1;\ 1/(1-z)) \quad (\beta - \alpha \neq 1, 2, 3, \dots) \tag{8.61}$$

and

$$(1-z)^{-\beta}\,{}_2F_1(\beta,\ \gamma-\alpha;\ \beta-\alpha+1;\ 1/(1-z)) \quad (\beta - \alpha \neq 0, -1, -2, \dots). \tag{8.62}$$

By (8.49), the solution (8.61) coincides with (8.47) up to the factor $(-1)^{\alpha}$, and (8.62) coincides with (8.48) up to the factor $(-1)^{\beta}$.

Altogether, through the three transformations in the first row of (8.42) we have obtained the following three pairs of solutions of (8.3):

$$\begin{cases} {}_2F_1(\alpha, \beta; \gamma; z) \quad (\gamma \neq 0, -1, -2, \dots) \\ z^{1-\gamma}\,{}_2F_1\,(\alpha-\gamma+1,\ \beta-\gamma+1;\ 2-\gamma;\ z) \quad (\gamma \neq 1, 2, 3, \dots) \end{cases} \tag{8.63}$$

for $z \in \mathbb{C}\backslash[1, +\infty)$, with (8.63) linearly independent if $\gamma \notin \mathbb{Z}$;

$$\begin{cases} {}_2F_1(\alpha,\ \beta;\ \alpha+\beta-\gamma+1;\ 1-z) \quad (\gamma-\alpha-\beta \neq 1, 2, 3, \dots) \\ (1-z)^{\gamma-\alpha-\beta}\,{}_2F_1(\gamma-\alpha,\ \gamma-\beta;\ \gamma-\alpha-\beta+1;\ 1-z) \\ \qquad (\gamma-\alpha-\beta \neq 0, -1, -2, \dots) \end{cases} \tag{8.64}$$

for $z \in \mathbb{C}\backslash(-\infty, 0]$, with (8.64) linearly independent if $\gamma - \alpha - \beta \notin \mathbb{Z}$;

$$\begin{cases} z^{-\alpha}\,{}_2F_1(\alpha,\ \alpha-\gamma+1;\ \alpha-\beta+1;\ 1/z) \quad (\beta-\alpha \neq 1, 2, 3, \dots) \\ z^{-\beta}\,{}_2F_1(\beta,\ \beta-\gamma+1;\ \beta-\alpha+1;\ 1/z) \quad (\beta-\alpha \neq 0, -1, -2, \dots) \end{cases} \tag{8.65}$$

for $z \in \mathbb{C}\backslash[0,1]$, with (8.65) linearly independent if $\beta - \alpha \notin \mathbb{Z}$.

If $\gamma \in \mathbb{Z}$, $\gamma - \alpha - \beta \in \mathbb{Z}$ and $\beta - \alpha \in \mathbb{Z}$, the general solution of (8.3) cannot be expressed as a linear combination of (8.63), (8.64) or (8.65). Thus, if γ, $\alpha + \beta$, $\alpha - \beta \in \mathbb{Z}$, a function of logarithmic type (7.73) cannot be avoided in the general solution of (8.3).

We conclude this discussion by giving an application of (8.50). Let

$$\alpha, \beta, \gamma \neq 0, -1, -2, \ldots \quad \text{and} \quad \operatorname{Re}(\gamma - \alpha - \beta) < 0. \tag{8.66}$$

Then

$$\lim_{z \to 1-} \left|(1-z)^{\gamma-\alpha-\beta}\right| = \lim_{z \to 1-} (1-z)^{\operatorname{Re}(\gamma-\alpha-\beta)} = +\infty.$$

Moreover $\operatorname{Re}(\gamma - (\gamma-\alpha) - (\gamma-\beta)) = \operatorname{Re}(\alpha+\beta-\gamma) > 0$ whence, by (8.30),

$$\lim_{z \to 1-} {}_2F_1(\gamma-\alpha, \gamma-\beta; \gamma; z) = \frac{\Gamma(\gamma)\,\Gamma(\alpha+\beta-\gamma)}{\Gamma(\alpha)\,\Gamma(\beta)} \neq 0.$$

Therefore, as a complement to Theorem 8.2, from the identity (8.50) we get

$$\lim_{z \to 1-} {}_2F_1(\alpha, \beta; \gamma; z) = \frac{\Gamma(\gamma)\,\Gamma(\alpha+\beta-\gamma)}{\Gamma(\alpha)\,\Gamma(\beta)} \lim_{z \to 1-} (1-z)^{\gamma-\alpha-\beta} = \infty,$$

under the assumptions (8.66).

With the next theorem we express ${}_2F_1(\alpha, \beta; \gamma; z)$ as a linear combination of the functions (8.64), or of (8.58) and (8.60), if $\gamma - \alpha - \beta \notin \mathbb{Z}$; of (8.65), or of (8.61) and (8.62), if $\beta - \alpha \notin \mathbb{Z}$.

Theorem 8.5 *Let $\alpha, \beta, \gamma \in \mathbb{C}$ with $\gamma \neq 0, -1, -2, \ldots$. If $\gamma - \alpha - \beta \notin \mathbb{Z}$ then, for any $z \in \mathbb{C}\backslash\big((-\infty, 0] \cup [1, +\infty)\big)$,*

$$\begin{aligned}
{}_2F_1(\alpha, \beta; \gamma; z) &= \frac{\Gamma(\gamma)\,\Gamma(\gamma-\alpha-\beta)}{\Gamma(\gamma-\alpha)\,\Gamma(\gamma-\beta)}\, {}_2F_1(\alpha, \beta; \alpha+\beta-\gamma+1; 1-z) \\
&+ \frac{\Gamma(\gamma)\,\Gamma(\alpha+\beta-\gamma)}{\Gamma(\alpha)\,\Gamma(\beta)} (1-z)^{\gamma-\alpha-\beta}\, {}_2F_1(\gamma-\alpha, \gamma-\beta; \gamma-\alpha-\beta+1; 1-z) \\
&= \frac{\Gamma(\gamma)\,\Gamma(\gamma-\alpha-\beta)}{\Gamma(\gamma-\alpha)\,\Gamma(\gamma-\beta)} z^{-\alpha}\, {}_2F_1(\alpha, \alpha-\gamma+1; \alpha+\beta-\gamma+1; (z-1)/z) \\
&+ \frac{\Gamma(\gamma)\,\Gamma(\alpha+\beta-\gamma)}{\Gamma(\alpha)\,\Gamma(\beta)} z^{\alpha-\gamma}(1-z)^{\gamma-\alpha-\beta} \\
&\times\ {}_2F_1(\gamma-\alpha, 1-\alpha; \gamma-\alpha-\beta+1; (z-1)/z).
\end{aligned} \tag{8.67}$$

If $\beta - \alpha \notin \mathbb{Z}$ then, for any $z \in \mathbb{C}\backslash[0, +\infty)$,

$$\begin{aligned}
{}_2F_1(\alpha,\beta; \gamma; z) &= \frac{\Gamma(\gamma)\,\Gamma(\beta-\alpha)}{\Gamma(\gamma-\alpha)\,\Gamma(\beta)} (-z)^{-\alpha}\, {}_2F_1(\alpha, \alpha-\gamma+1; \alpha-\beta+1; 1/z) \\
&+ \frac{\Gamma(\gamma)\,\Gamma(\alpha-\beta)}{\Gamma(\alpha)\,\Gamma(\gamma-\beta)} (-z)^{-\beta}\, {}_2F_1(\beta, \beta-\gamma+1; \beta-\alpha+1; 1/z) \\
&= \frac{\Gamma(\gamma)\,\Gamma(\beta-\alpha)}{\Gamma(\gamma-\alpha)\,\Gamma(\beta)} (1-z)^{-\alpha}\, {}_2F_1(\alpha, \gamma-\beta; \alpha-\beta+1; 1/(1-z))
\end{aligned} \tag{8.68}$$

$$+\frac{\Gamma(\gamma)\,\Gamma(\alpha-\beta)}{\Gamma(\alpha)\,\Gamma(\gamma-\beta)}(1-z)^{-\beta}\,{}_2F_1(\beta,\,\gamma-\alpha;\,\beta-\alpha+1;\,1/(1-z)).$$

Proof If $\gamma-\alpha-\beta\notin\mathbb{Z}$, the general solution of the hypergeometric differential equation (8.3) is a linear combination of the functions (8.64) with constant coefficients (i.e., independent of z). In particular, since $\gamma\neq 0,-1,-2,\ldots$, there exist coefficients $C_1=C_1(\alpha,\beta,\gamma)$ and $C_2=C_2(\alpha,\beta,\gamma)$ such that

$$\begin{aligned}{}_2F_1(\alpha,\beta;\gamma;z)&=C_1\,{}_2F_1(\alpha,\,\beta;\,\alpha+\beta-\gamma+1;\,1-z)\\&\quad+C_2(1-z)^{\gamma-\alpha-\beta}\,{}_2F_1(\gamma-\alpha,\,\gamma-\beta;\,\gamma-\alpha-\beta+1;\,1-z)\end{aligned}\tag{8.69}$$

for all

$$z\in\mathbb{C}\backslash\big((-\infty,0]\cup[1,+\infty)\big).\tag{8.70}$$

In order to find C_1 and C_2, we temporarily assume $\operatorname{Re}(\alpha+\beta)<\operatorname{Re}\gamma<1$. We take the limits of (8.69) for $z\to 1-$ and for $z\to 0+$. Since

$$\lim_{z\to 1-}\left|(1-z)^{\gamma-\alpha-\beta}\right|=\lim_{z\to 1-}(1-z)^{\operatorname{Re}(\gamma-\alpha-\beta)}=0,$$

by (8.69) and by Theorem 8.2 we obtain

$$C_1=\lim_{z\to 1-}{}_2F_1(\alpha,\beta;\gamma;z)=\frac{\Gamma(\gamma)\,\Gamma(\gamma-\alpha-\beta)}{\Gamma(\gamma-\alpha)\,\Gamma(\gamma-\beta)}\tag{8.71}$$

and

$$C_1\frac{\Gamma(\alpha+\beta-\gamma+1)\,\Gamma(1-\gamma)}{\Gamma(\beta-\gamma+1)\,\Gamma(\alpha-\gamma+1)}+C_2\frac{\Gamma(\gamma-\alpha-\beta+1)\,\Gamma(1-\gamma)}{\Gamma(1-\alpha)\,\Gamma(1-\beta)}=\lim_{z\to 0+}{}_2F_1(\alpha,\beta;\gamma;z)=1.$$

Hence, by (8.71),

$$\begin{aligned}C_2&=\frac{\Gamma(1-\alpha)\,\Gamma(1-\beta)}{\Gamma(\gamma-\alpha-\beta+1)\,\Gamma(1-\gamma)}\\&\quad\times\left(1-\frac{\Gamma(\gamma)\,\Gamma(\gamma-\alpha-\beta)}{\Gamma(\gamma-\alpha)\,\Gamma(\gamma-\beta)}\cdot\frac{\Gamma(\alpha+\beta-\gamma+1)\,\Gamma(1-\gamma)}{\Gamma(\alpha-\gamma+1)\,\Gamma(\beta-\gamma+1)}\right).\end{aligned}\tag{8.72}$$

With the values (8.71) and (8.72) for C_1 and C_2, we can apply Theorem 8.3 to (8.69). By analytic continuation on α, β and γ, we can drop the restrictive assumption $\operatorname{Re}(\alpha+\beta)<\operatorname{Re}\gamma<1$. Therefore, for any fixed z satisfying (8.70), the identity (8.69) holds for all α, β and γ such that $\gamma\neq 0,-1,-2,\ldots$ and $\gamma-\alpha-\beta\notin\mathbb{Z}$.

In order to express C_2 by a formula more satisfactory than (8.72), we apply (8.50) to the left-hand side of (8.69). Dividing by $(1-z)^{\gamma-\alpha-\beta}$ we obtain

$$\begin{aligned}{}_2F_1(\gamma-\alpha,\,\gamma-\beta;\,\gamma;\,z) &= C_1(1-z)^{\alpha+\beta-\gamma}\,{}_2F_1(\alpha,\,\beta;\,\alpha+\beta-\gamma+1;\,1-z)\\ &\quad + C_2\,{}_2F_1(\gamma-\alpha,\,\gamma-\beta;\,\gamma-\alpha-\beta+1;\,1-z).\end{aligned}$$

Assuming now $\mathrm{Re}(\alpha+\beta-\gamma)>0$ and making $z\to 1-$, by Theorem 8.2 we get

$$C_2 = \lim_{z\to 1-}\,{}_2F_1(\gamma-\alpha,\,\gamma-\beta;\,\gamma;\,z) = \frac{\Gamma(\gamma)\,\Gamma(\alpha+\beta-\gamma)}{\Gamma(\alpha)\,\Gamma(\beta)}. \tag{8.73}$$

As before, by analytic continuation on α, β and γ, we can drop the restriction $\mathrm{Re}(\alpha+\beta-\gamma)>0$. Then (8.67) follows from (8.69), (8.71) and (8.73), and by recalling that (8.58) coincides with (8.44), and (8.60) coincides with (8.45) up to the factor $(-1)^{\gamma-\alpha-\beta}$.

If $\beta-\alpha\notin\mathbb{Z}$ and $z\in\mathbb{C}\setminus[\,0,+\infty)$, we apply first (8.49) and then (8.67). We get

$$\begin{aligned}{}_2F_1(\alpha,\beta;\gamma;z) &= \frac{\Gamma(\gamma)\,\Gamma(\beta-\alpha)}{\Gamma(\gamma-\alpha)\,\Gamma(\beta)}(1-z)^{-\alpha}\,{}_2F_1(\alpha,\,\gamma-\beta;\,\alpha-\beta+1;\,1/(1-z))\\ &\quad+\frac{\Gamma(\gamma)\,\Gamma(\alpha-\beta)}{\Gamma(\alpha)\,\Gamma(\gamma-\beta)}(1-z)^{-\beta}\,{}_2F_1(\gamma-\alpha,\,\beta;\,\beta-\alpha+1;\,1/(1-z)).\end{aligned} \tag{8.74}$$

But, as we have shown, (8.61) coincides with (8.47) up to the factor $(-1)^{\alpha}$, and (8.62) coincides with (8.48) up to the factor $(-1)^{\beta}$. Therefore (8.68) follows from (8.74). □

8.6 The Confluent Hypergeometric Function ${}_1F_1$

A natural generalization of the hypergeometric function (8.12) is

$$ {}_pF_s(\alpha_1,\ldots,\alpha_p;\,\gamma_1,\ldots,\gamma_s;\,z) = \sum_{n=0}^{\infty}\frac{(\alpha_1)_n\cdots(\alpha_p)_n}{(\gamma_1)_n\cdots(\gamma_s)_n}\,\frac{z^n}{n!} \tag{8.75}$$

$$(\gamma_k\neq 0,-1,-2,\ldots;\ k=1,\ldots,s)$$

for any $p,s\in\mathbb{N}$, where $\alpha_1,\ldots,\alpha_p;\ \gamma_1,\ldots,\gamma_s$ are complex parameters and $(\alpha_h)_n$, $(\gamma_k)_n$ are Pochhammer symbols defined in (8.11).

If $-\alpha_h\in\mathbb{N}$ for at least one h $(1\le h\le p)$, (8.75) is plainly a polynomial in z. If $\alpha_h\neq 0,-1,-2,\ldots$ for $h=1,\ldots,p$, (8.75) is a Taylor series with radius of convergence ϱ such that

$$\varrho = \begin{cases} +\infty, & \text{if } p \le s \\ 1, & \text{if } p = s+1 \\ 0, & \text{if } p > s+1 \end{cases} \tag{8.76}$$

because

$$\begin{aligned} \varrho &= \lim_{n\to\infty} \left| \frac{(\alpha_1)_n \cdots (\alpha_p)_n}{(\gamma_1)_n \cdots (\gamma_s)_n \, n!} \cdot \frac{(\gamma_1)_{n+1} \cdots (\gamma_s)_{n+1} \, (n+1)!}{(\alpha_1)_{n+1} \cdots (\alpha_p)_{n+1}} \right| \\ &= \lim_{n\to\infty} \left| \frac{(\gamma_1+n)\cdots(\gamma_s+n)(n+1)}{(\alpha_1+n)\cdots(\alpha_p+n)} \right| = \lim_{n\to\infty} n^{s-p+1}. \end{aligned}$$

Beside the hypergeometric function ${}_2F_1$, of special interest is the *confluent hypergeometric function* ${}_1F_1(\alpha; \gamma; z)$, first introduced by Kummer and also commonly denoted by $\Phi(\alpha; \gamma; z)$. Thus

$$\Phi(\alpha; \gamma; z) = {}_1F_1(\alpha; \gamma; z) = \sum_{n=0}^{\infty} \frac{(\alpha)_n}{(\gamma)_n} \frac{z^n}{n!} \qquad (\gamma \ne 0, -1, -2, \dots). \tag{8.77}$$

If $\alpha = -N$ with $N \in \mathbb{N}$, $\Phi(\alpha; \gamma; z)$ is a polynomial in z of degree N. If $\alpha \ne 0, -1, -2, \dots$, by (8.76) $\Phi(\alpha; \gamma; z)$ is the sum of a Taylor series with radius of convergence $+\infty$. Thus in either case $\Phi(\alpha; \gamma; z) = {}_1F_1(\alpha; \gamma; z)$ is an entire function of z.

Similarly to Theorem 8.3, for any fixed $z \in \mathbb{C}$ the function

$$\Phi^*(\alpha; \gamma; z) := \frac{\Phi(\alpha; \gamma; z)}{\Gamma(\gamma)} = \frac{{}_1F_1(\alpha; \gamma; z)}{\Gamma(\gamma)}$$

is an entire function of α and γ. For, by (8.17),

$$\Phi^*(\alpha; \gamma; z) = \frac{\Phi(\alpha; \gamma; z)}{\Gamma(\gamma)} = \sum_{n=0}^{\infty} \frac{(\alpha)_n}{\Gamma(\gamma+n)} \frac{z^n}{n!}. \tag{8.78}$$

By (6.29), $1/\Gamma(\gamma+n)$ is an entire function of γ. Hence each term in the series (8.78) is a polynomial in α and an entire function of γ, and the series is totally convergent, and hence uniformly convergent, for $|\alpha| \le A$ and $|\gamma| \le C$, where the constants A and C can be taken arbitrarily large, because, for any sufficiently large n,

$$\begin{aligned} \left| \frac{(\alpha)_{n+1}}{\Gamma(\gamma+n+1)} \frac{z^{n+1}}{(n+1)!} \cdot \frac{\Gamma(\gamma+n)}{(\alpha)_n} \frac{n!}{z^n} \right| &= \left| \frac{\alpha+n}{\gamma+n} \frac{z}{n+1} \right| \\ &\le \frac{A+n}{(n-C)(n+1)} |z| < \frac{1}{2}, \end{aligned}$$

whence

$$\sum_n \left| \frac{(\alpha)_n}{\Gamma(\gamma+n)} \frac{z^n}{n!} \right| \ll \sum_n (1/2)^n < +\infty.$$

Since $\Phi^*(\alpha;\gamma;z)$ in (8.78) is an entire function of γ, it is defined, unlike $\Phi(\alpha;\gamma;z)$, also for $\gamma = 0, -1, -2, \ldots$. Since $1/\Gamma(\gamma+n)$ vanishes at

$$\gamma = -n,\ -(n+1),\ -(n+2),\ \ldots,$$

(8.78) yields, for $s = 1, 2, 3, \ldots$,

$$\begin{aligned}\Phi^*(\alpha;\ 1-s;\ z) &= \sum_{n=s}^{\infty} \frac{(\alpha)_n}{\Gamma(n-s+1)} \frac{z^n}{n!} \\ &= \sum_{n=s}^{\infty} \frac{(\alpha)_n}{(n-s)!} \frac{z^n}{n!} = \sum_{\nu=0}^{\infty} \frac{(\alpha)_s\,(\alpha+s)_\nu}{\nu!\,(s+\nu)!} z^{s+\nu},\end{aligned}$$

whence

$$\Phi^*(\alpha;\ 1-s;\ z) = (\alpha)_s\, z^s\, \Phi^*(\alpha+s;\ s+1;\ z) \quad (s = 1, 2, 3, \ldots). \tag{8.79}$$

Other elementary properties of Φ are similar to the corresponding properties of ${}_2F_1$. For instance, as in (8.13), we plainly have the differentiation formula

$$\frac{\mathrm{d}^m}{\mathrm{d}z^m}\,\Phi(\alpha;\gamma;z) = \frac{(\alpha)_m}{(\gamma)_m}\,\Phi(\alpha+m;\ \gamma+m;\ z) \quad (m = 1, 2, 3, \ldots). \tag{8.80}$$

Also, the analogues of Gauss' contiguity formulae for ${}_2F_1$ hold for Φ. We abbreviate

$$\Phi(\alpha;\gamma;z) = \Phi, \quad \Phi(\alpha\pm 1;\gamma;z) = \Phi(\alpha\pm 1), \quad \Phi(\alpha;\gamma\pm 1;z) = \Phi(\gamma\pm 1).$$

By direct substitution of the series (8.77) one easily proves the following $\binom{4}{2} = 6$ contiguity formulae:

$$\begin{aligned}&(2\alpha-\gamma+z)\Phi - \alpha\,\Phi(\alpha+1) + (\gamma-\alpha)\Phi(\alpha-1) = 0,\\ &\gamma(\alpha+z)\Phi - \alpha\gamma\,\Phi(\alpha+1) - (\gamma-\alpha)z\,\Phi(\gamma+1) = 0,\\ &(\alpha-\gamma+1)\Phi - \alpha\,\Phi(\alpha+1) + (\gamma-1)\Phi(\gamma-1) = 0,\\ &\gamma\,\Phi - \gamma\,\Phi(\alpha-1) - z\,\Phi(\gamma+1) = 0,\\ &(\alpha+z-1)\Phi + (\gamma-\alpha)\Phi(\alpha-1) - (\gamma-1)\Phi(\gamma-1) = 0,\\ &\gamma(\gamma+z-1)\Phi - (\gamma-\alpha)z\,\Phi(\gamma+1) - \gamma(\gamma-1)\Phi(\gamma-1) = 0.\end{aligned} \tag{8.81}$$

The analogue of Euler's integral representation (8.16) of ${}_2F_1$ is given by the following

Theorem 8.6 *If* $\operatorname{Re}\gamma > \operatorname{Re}\alpha > 0$, *for any* $z \in \mathbb{C}$ *we have*

$$\Phi(\alpha;\gamma;z) = \frac{\Gamma(\gamma)}{\Gamma(\alpha)\Gamma(\gamma-\alpha)} \int_0^1 e^{zt}\, t^{\alpha-1}(1-t)^{\gamma-\alpha-1}\, dt. \tag{8.82}$$

Proof We substitute (8.18) in (8.77). As in the proof of (8.19) we get

$$\begin{aligned}
\Phi(\alpha;\gamma;z) &= \frac{\Gamma(\gamma)}{\Gamma(\alpha)\Gamma(\gamma-\alpha)} \sum_{n=0}^{\infty} \frac{z^n}{n!} \int_0^1 t^{\alpha-1+n}(1-t)^{\gamma-\alpha-1}\, dt \\
&= \frac{\Gamma(\gamma)}{\Gamma(\alpha)\Gamma(\gamma-\alpha)} \int_0^1 \sum_{n=0}^{\infty} \frac{(zt)^n}{n!}\, t^{\alpha-1}(1-t)^{\gamma-\alpha-1}\, dt \\
&= \frac{\Gamma(\gamma)}{\Gamma(\alpha)\Gamma(\gamma-\alpha)} \int_0^1 e^{zt}\, t^{\alpha-1}(1-t)^{\gamma-\alpha-1}\, dt.
\end{aligned}$$

□

As a consequence of (8.82) we obtain Kummer's transformation formula (8.83) below.

Theorem 8.7 (Kummer) *For any* $\alpha, \gamma, z \in \mathbb{C}$ *with* $\gamma \neq 0, -1, -2, \ldots,$

$$\Phi(\alpha;\gamma;z) = e^z\, \Phi(\gamma-\alpha;\gamma;-z). \tag{8.83}$$

Proof If $\operatorname{Re}\gamma > \operatorname{Re}\alpha > 0$ we substitute $t = 1 - T$ in the integral (8.82). We get

$$\begin{aligned}
\Phi(\alpha;\gamma;z) &= e^z \frac{\Gamma(\gamma)}{\Gamma(\alpha)\Gamma(\gamma-\alpha)} \int_0^1 e^{-zT}\, T^{\gamma-\alpha-1}(1-T)^{\alpha-1}\, dT \\
&= e^z\, \Phi(\gamma-\alpha;\gamma;-z).
\end{aligned}$$

Since both sides of (8.83) divided by $\Gamma(\gamma)$ are entire functions of α and γ, by analytic continuation on α and γ we can drop the restrictive assumption $\operatorname{Re}\gamma > \operatorname{Re}\alpha > 0$. Hence (8.83) holds for any $\alpha, \gamma, z \in \mathbb{C}$ with $\gamma \neq 0, -1, -2, \ldots$. □

8.7 The Confluent Hypergeometric Equation

In this section we discuss the confluent hypergeometric function $\Phi(\alpha;\gamma;z) = {}_1F_1(\alpha;\gamma;z)$ as a solution of the confluent hypergeometric differential equation.

Since

$$ {}_2F_1(\alpha, \beta; \gamma; z/\beta) = 1 + \sum_{n=1}^{\infty} \frac{(\alpha)_n}{(\gamma)_n} \left(1 + \frac{1}{\beta}\right) \cdots \left(1 + \frac{n-1}{\beta}\right) \frac{z^n}{n!}, $$

we get

$$ \Phi(\alpha; \gamma; z) = \lim_{\beta \to \infty} {}_2F_1(\alpha, \beta; \gamma; z/\beta). \tag{8.84} $$

The hypergeometric function ${}_2F_1(\alpha, \beta; \gamma; z/\beta)$ satisfies the differential equation (8.3) with z replaced by z/β, i.e.,

$$ z\left(1 - \frac{z}{\beta}\right) \frac{\mathrm{d}^2 w}{\mathrm{d}z^2} + \left(\gamma - \frac{\alpha + \beta + 1}{\beta} z\right) \frac{\mathrm{d}w}{\mathrm{d}z} - \alpha w = 0, \tag{8.85} $$

with regular singular points at $z = 0, \beta, \infty$. Making $\beta \to \infty$ in (8.85), we get the *confluent hypergeometric differential equation*:

$$ z\, w'' + (\gamma - z)\, w' - \alpha w = 0. \tag{8.86} $$

Hence (8.86) can be viewed as the limit case of (8.85) at the 'confluence' of the fuchsian points β and ∞ (whence the denomination).

Writing (8.86) in the form (8.1), the coefficients $p_1(z)$ and $p_2(z)$ are

$$ p_1(z) = \frac{\gamma - z}{z}, \quad p_2(z) = -\frac{\alpha}{z}. $$

Thus $z = 0$ is a regular singular point of (8.86), while Fuchs' conditions are not satisfied at the confluence point $z = \infty$. As we did in Sect. 8.2 for the hypergeometric differential equation (8.3), we seek solutions of (8.86) of the form (7.42) with $z_0 = 0$, i.e., in the present case, for $0 < |z| < +\infty$. With the notation (7.48), (7.49) we have

$$ A(z) = \gamma - z, \quad B(z) = -\alpha z, $$

whence

$$ \begin{aligned} a_0 &= \gamma,\ a_1 = -1,\ a_2 = a_3 = \ldots = 0, \\ b_0 &= 0,\ b_1 = -\alpha,\ b_2 = b_3 = \ldots = 0. \end{aligned} $$

Thus the auxiliary functions (7.54) are

$$ \begin{aligned} f_0(t) &= t(t + \gamma - 1), \\ f_1(t) &= -(t + \alpha), \\ f_\nu(t) &= 0 \quad (\nu = 2, 3, \ldots), \end{aligned} \tag{8.87} $$

and the indicial equation is again (8.5). Therefore the roots of the indicial equation are

$$\begin{aligned} r_1 &= 0 \text{ and } r_2 = 1-\gamma, \text{ if } \operatorname{Re}\gamma \geq 1, \\ r_1 &= 1-\gamma \text{ and } r_2 = 0, \text{ if } \operatorname{Re}\gamma < 1. \end{aligned}$$

Hence if $\gamma \notin \mathbb{Z}$, (8.86) has two linearly independent solutions of the form

$$\sum_{n=0}^{\infty} c_n z^n \text{ with } c_0 = 1 \tag{8.88}$$

and

$$z^{1-\gamma} \sum_{n=0}^{\infty} \widetilde{c}_n z^n \text{ with } \widetilde{c}_0 = 1, \tag{8.89}$$

with radii of convergence $+\infty$. If $\gamma \in \mathbb{Z}$, we have the solution (8.88) if $\gamma = 1, 2, 3, \ldots$, or the solution (8.89) if $\gamma = 0, -1, -2, \ldots$.

Let $\gamma \in \mathbb{C} \backslash \{0, -1, -2, \ldots\}$. By (8.87), the recursion (7.55) with $r = 0$ is now

$$c_n = \frac{n+\alpha-1}{n(n+\gamma-1)} c_{n-1} \quad (n = 1, 2, 3, \ldots). \tag{8.90}$$

From $c_0 = 1$ we get, by induction on n,

$$c_n = \frac{(\alpha)_n}{(\gamma)_n\, n!} \quad (n = 0, 1, 2, \ldots), \tag{8.91}$$

with Pochhammer symbols $(\alpha)_n$ and $(\gamma)_n$. Hence the solution (8.88) of the differential equation (8.86) is the confluent hypergeometric function $\Phi(\alpha; \gamma; z)$ in (8.77), as expected by (8.84).

If $\gamma \in \mathbb{C} \backslash \{1, 2, 3, \ldots\}$, by (8.87) the recursion (7.55) with $r = 1 - \gamma$ is

$$\widetilde{c}_n = \frac{n+\alpha-\gamma}{n(n-\gamma+1)} \widetilde{c}_{n-1} \quad (n = 1, 2, 3, \ldots). \tag{8.92}$$

With the involutory substitution

$$\begin{cases} \alpha = \widetilde{\alpha} - \widetilde{\gamma} + 1 \\ \gamma = 2 - \widetilde{\gamma} \end{cases}$$

(8.92) becomes (8.90), with α, γ, c replaced by $\widetilde{\alpha}, \widetilde{\gamma}, \widetilde{c}$ respectively. Hence (8.91) yields

$$\widetilde{c}_n = \frac{(\widetilde{\alpha})_n}{(\widetilde{\gamma})_n\, n!} = \frac{(\alpha-\gamma+1)_n}{(2-\gamma)_n\, n!} \qquad (n = 0, 1, 2, \dots).$$

Thus the solution (8.89) is

$$z^{1-\gamma}\, \Phi(\alpha-\gamma+1;\ 2-\gamma;\ z) \qquad (\gamma \neq 1, 2, 3, \dots). \tag{8.93}$$

Therefore, if $\gamma \notin \mathbb{Z}$ the general solution of (8.86) for $z \in \mathbb{C}\backslash\{0\}$ is

$$C_1\, \Phi(\alpha; \gamma; z) + C_2\, z^{1-\gamma}\, \Phi(\alpha-\gamma+1;\ 2-\gamma;\ z), \tag{8.94}$$

with arbitrary constants C_1 and C_2.

We have seen that the function

$$\Phi^*(\alpha; \gamma; z) = \frac{\Phi(\alpha; \gamma; z)}{\Gamma(\gamma)}$$

in (8.78) is an entire function of γ. Hence, by analytic continuation on γ, $\Phi^*(\alpha; \gamma; z)$ is a solution of (8.86) also for $\gamma = 0, -1, -2, \dots$, and hence for all $\alpha, \gamma \in \mathbb{C}$. Similarly,

$$z^{1-\gamma}\, \Phi^*(\alpha-\gamma+1;\ 2-\gamma;\ z) = z^{1-\gamma}\, \frac{\Phi(\alpha-\gamma+1;\ 2-\gamma;\ z)}{\Gamma(2-\gamma)}$$

is a solution of (8.86) also for $\gamma = 1, 2, 3, \dots$, hence for all $\alpha, \gamma \in \mathbb{C}$. However, as we shall now prove, the general solution of (8.86) can be written in the form

$$C_1\, \Phi^*(\alpha; \gamma; z) + C_2\, z^{1-\gamma}\, \Phi^*(\alpha-\gamma+1;\ 2-\gamma;\ z)$$

with arbitrary constants C_1 and C_2, if and only if $\gamma \notin \mathbb{Z}$.

Using the differentiation formula (8.80) for $m = 1$, it is easy to compute the wronskian of the solutions $\Phi^*(\alpha; \gamma; z)$ and $z^{1-\gamma}\, \Phi^*(\alpha-\gamma+1;\ 2-\gamma;\ z)$ of (8.86). We recall that if w_1 and w_2 are any two solutions of the differential equation (8.1), the wronskian of w_1 and w_2 is given by Liouville's formula

$$W = W(w_1, w_2) = \begin{vmatrix} w_1 & w_2 \\ w_1' & w_2' \end{vmatrix} = C \exp\left(-\textstyle\int p_1(z)\, \mathrm{d}z\right),$$

where C is a constant (i.e., depending on w_1 and w_2, but independent of z). Liouville's formula is easily proved by remarking that

$$\frac{\mathrm{d}W}{\mathrm{d}z} = \begin{vmatrix} w_1 & w_2 \\ w_1'' & w_2'' \end{vmatrix} = -p_1(z) W.$$

In particular, the wronskian of any two solutions of (8.86) is

$$C \exp \int \left(1 - \frac{\gamma}{z}\right) \mathrm{d}z = C e^z z^{-\gamma}. \tag{8.95}$$

By (8.80),

$$\begin{aligned} \frac{\mathrm{d}}{\mathrm{d}z} \Phi^*(\alpha; \gamma; z) &= \frac{1}{\Gamma(\gamma)} \frac{\mathrm{d}}{\mathrm{d}z} \Phi(\alpha; \gamma; z) \\ &= \frac{\alpha}{\gamma \Gamma(\gamma)} \Phi(\alpha + 1; \gamma + 1; z) = \alpha \Phi^*(\alpha + 1; \gamma + 1; z). \end{aligned} \tag{8.96}$$

This easily yields

$$\begin{aligned} &W\big(\Phi^*(\alpha; \gamma; z), z^{1-\gamma} \Phi^*(\alpha - \gamma + 1; 2 - \gamma; z)\big) \\ &\quad = (1 - \gamma) z^{-\gamma} \Phi^*(\alpha; \gamma; z) \Phi^*(\alpha - \gamma + 1; 2 - \gamma; z) \\ &\qquad + z^{1-\gamma}\big((\alpha - \gamma + 1) \Phi^*(\alpha; \gamma; z) \Phi^*(\alpha - \gamma + 2; 3 - \gamma; z) \\ &\qquad\qquad - \alpha \Phi^*(\alpha + 1; \gamma + 1; z) \Phi^*(\alpha - \gamma + 1; 2 - \gamma; z)\big) \\ &\quad = C e^z z^{-\gamma} \end{aligned}$$

by (8.95). Multiplying by z^γ and then putting $z = 0$ we obtain

$$\begin{aligned} C &= (1 - \gamma) \Phi^*(\alpha; \gamma; 0) \Phi^*(\alpha - \gamma + 1; 2 - \gamma; 0) \\ &= (1 - \gamma) \frac{1}{\Gamma(\gamma)} \frac{1}{\Gamma(2 - \gamma)} = \frac{1}{\Gamma(\gamma)\Gamma(1 - \gamma)} = \frac{\sin(\pi\gamma)}{\pi}, \end{aligned}$$

where we have used Euler's reflection formula (6.21). It follows that

$$W\big(\Phi^*(\alpha; \gamma; z), z^{1-\gamma} \Phi^*(\alpha - \gamma + 1; 2 - \gamma; z)\big) = \frac{\sin(\pi\gamma)}{\pi} e^z z^{-\gamma},$$

which shows that $\Phi^*(\alpha; \gamma; z)$ and $z^{1-\gamma} \Phi^*(\alpha - \gamma + 1; 2 - \gamma; z)$ are linearly independent if and only if $\gamma \notin \mathbb{Z}$.

In order to treat the case $\gamma \in \mathbb{Z}$, it is convenient to introduce a new function $\Psi(\alpha; \gamma; z)$ which is a solution of (8.86) for all $\alpha, \gamma \in \mathbb{C}$. Also, as we shall prove, the general solution of (8.86) is

$$C_1 \Phi^*(\alpha; \gamma; z) + C_2 \Psi(\alpha; \gamma; z)$$

with arbitrary constants C_1 and C_2, if and only if $\alpha \neq 0, -1, -2, \ldots$

If $\gamma \notin \mathbb{Z}$ we define, for any $\alpha \in \mathbb{C}$ and any $z \in \mathbb{C}\backslash\{0\}$,

$$\begin{aligned}\Psi(\alpha;\gamma;z) &= \frac{\Gamma(1-\gamma)}{\Gamma(\alpha-\gamma+1)}\,\Phi(\alpha;\gamma;z) \\ &\quad + \frac{\Gamma(\gamma-1)}{\Gamma(\alpha)}\,z^{1-\gamma}\,\Phi(\alpha-\gamma+1;\,2-\gamma;\,z) \\ &= \frac{\Gamma(\gamma)\Gamma(1-\gamma)}{\Gamma(\alpha-\gamma+1)}\,\Phi^*(\alpha;\gamma;z) \\ &\quad + \frac{\Gamma(\gamma-1)\Gamma(2-\gamma)}{\Gamma(\alpha)}\,z^{1-\gamma}\,\Phi^*(\alpha-\gamma+1;\,2-\gamma;\,z).\end{aligned} \tag{8.97}$$

$\Psi(\alpha;\gamma;z)$ is called the *confluent hypergeometric function of the second kind*. Since $\Gamma(\gamma-1)/\Gamma(\alpha)=0$ if and only if $\alpha=0,-1,-2,\dots$, and $z^{1-\gamma}$ is a one-valued function of z if and only if $\gamma\in\mathbb{Z}$, from (8.97) we see that, for $\gamma\notin\mathbb{Z}$, $\Psi(\alpha;\gamma;z)$ is a multivalued function of $z\in\mathbb{C}\backslash\{0\}$ if and only if $\alpha\neq 0,-1,-2,\dots$. In this case, its principal value is defined by taking as usual $z^{1-\gamma}=\exp\big((1-\gamma)\log z\big)$ with $\log z=\log|z|+i\arg z$, $-\pi<\arg z\le\pi$, so that $\Psi(\alpha;\gamma;z)$ is one-valued and regular for z in the cut plane $\mathbb{C}\backslash\mathbb{R}^-$.

By (8.94), $\Psi(\alpha;\gamma;z)$ is a solution of the differential equation (8.86) when $\gamma\notin\mathbb{Z}$. Moreover, we shall show that $\Psi(\alpha;\gamma;z)$ takes finite values for $\gamma\in\mathbb{Z}$, so that, by analytic continuation, $\Psi(\alpha;\gamma;z)$ is a solution of (8.86) for all $\alpha,\gamma\in\mathbb{C}$.

From (8.97) we get, for any $\alpha\in\mathbb{C}$ and any $\gamma\in\mathbb{C}\backslash\mathbb{Z}$,

$$\begin{aligned}\Psi(\alpha;\gamma;z) &= \frac{\pi}{\sin(\pi\gamma)}\,\frac{\Phi^*(\alpha;\gamma;z)}{\Gamma(\alpha-\gamma+1)}+\frac{\pi}{\sin(\pi(\gamma-1))}\,z^{1-\gamma}\,\frac{\Phi^*(\alpha-\gamma+1;\,2-\gamma;\,z)}{\Gamma(\alpha)} \\ &= \frac{\pi}{\sin(\pi\gamma)}\left(\frac{\Phi^*(\alpha;\gamma;z)}{\Gamma(\alpha-\gamma+1)}-z^{1-\gamma}\,\frac{\Phi^*(\alpha-\gamma+1;\,2-\gamma;\,z)}{\Gamma(\alpha)}\right).\end{aligned} \tag{8.98}$$

First we extend $\Psi(\alpha;\gamma;z)$ for $\gamma=s+1$ $(s=0,1,2,\dots)$. If $\alpha=-m$ $(m=0,1,2,\dots)$, since $1/\Gamma(-m)=0$ and Φ^* is an entire function of the parameters, (8.98) yields

$$\Psi(-m;\gamma;z)=\frac{\pi\,\Phi^*(-m;\gamma;z)}{\sin(\pi\gamma)\,\Gamma(-m-\gamma+1)}. \tag{8.99}$$

From the Taylor expansion of $\sin(\pi\gamma)$ and the Laurent expansion of $\Gamma(-m-\gamma+1)$ around $\gamma=s+1$ $(s=0,1,2,\dots)$ we easily obtain, by (6.8),

$$\lim_{\gamma\to s+1}\sin(\pi\gamma)\,\Gamma(-m-\gamma+1)=\frac{(-1)^m\,\pi}{(m+s)!}.$$

Thus (8.99) yields, for $m = 0, 1, 2, \ldots$ and $s = 0, 1, 2, \ldots$,

$$\begin{aligned} \Psi(-m;\ s+1;\ z) &= (-1)^m\,(m+s)!\ \Phi^*(-m;\ s+1;\ z) \\ &= (-1)^m\,\frac{(m+s)!}{s!}\,\Phi(-m;\ s+1;\ z). \end{aligned} \tag{8.100}$$

We now assume $\alpha \neq 0, -1, -2, \ldots$. For $\gamma \notin \mathbb{Z}$, we replace in (8.98) $\Phi^*(\alpha;\gamma;z)$ and $\Phi^*(\alpha-\gamma+1;\ 2-\gamma;\ z)$ with the corresponding series (8.78). We get

$$\Psi(\alpha;\gamma;z) = \pi\,\frac{f(\alpha;\gamma;z) - g(\alpha;\gamma;z)}{\sin(\pi\gamma)}, \tag{8.101}$$

where

$$f(\alpha;\gamma;z) := \frac{1}{\Gamma(\alpha-\gamma+1)}\sum_{n=0}^{\infty}\frac{(\alpha)_n}{\Gamma(\gamma+n)}\,\frac{z^n}{n!} \tag{8.102}$$

and

$$g(\alpha;\gamma;z) := \frac{1}{\Gamma(\alpha)}\sum_{n=0}^{\infty}\frac{(\alpha-\gamma+1)_n}{\Gamma(2-\gamma+n)}\,\frac{z^{1-\gamma+n}}{n!}. \tag{8.103}$$

By (8.17),

$$\frac{(\alpha-s)_{s+n}}{\Gamma(\alpha)} = \frac{(\alpha)_n}{\Gamma(\alpha-s)}. \tag{8.104}$$

Plainly

$$\frac{1}{\Gamma(n-s+1)} = 0 \quad \text{for } n = 0, 1, \ldots, s-1. \tag{8.105}$$

By (8.102), (8.103), (8.104) and (8.105) we get

$$\begin{aligned} g(\alpha;\ s+1;\ z) &= \frac{1}{\Gamma(\alpha)}\sum_{n=s}^{\infty}\frac{(\alpha-s)_n\,z^{n-s}}{\Gamma(n-s+1)\,n!} = \frac{1}{\Gamma(\alpha)}\sum_{n=0}^{\infty}\frac{(\alpha-s)_{s+n}\,z^n}{n!\,(s+n)!} \\ &= \frac{1}{\Gamma(\alpha-s)}\sum_{n=0}^{\infty}\frac{(\alpha)_n\,z^n}{n!\,(s+n)!} = f(\alpha;\ s+1;\ z). \end{aligned}$$

Thus (8.101) yields, by L'Hôpital's rule,

$$\Psi(\alpha;\, s+1;\, z) = \lim_{\gamma\to s+1} \pi\, \frac{f(\alpha;\gamma;z) - g(\alpha;\gamma;z)}{\sin(\pi\gamma)} \qquad (8.106)$$

$$= (-1)^{s+1}\left[\frac{\partial f}{\partial\gamma} - \frac{\partial g}{\partial\gamma}\right]_{\gamma=s+1}.$$

As in Sect. 6.4, we denote by

$$\psi(t) = \frac{\Gamma'(t)}{\Gamma(t)}$$

the logarithmic derivative of the gamma-function. Then

$$\frac{\mathrm{d}}{\mathrm{d}t}\frac{1}{\Gamma(t)} = -\frac{\Gamma'(t)}{\Gamma(t)^2} = -\frac{\psi(t)}{\Gamma(t)}. \qquad (8.107)$$

As we remarked, (8.78) is a series of entire functions of γ, uniformly convergent for γ in any bounded subset of $\mathbb{C}$. Hence term-by-term differentiation with respect to γ is allowed. By (8.102) and (8.107) we get

$$\frac{\partial f}{\partial\gamma} = \frac{\psi(\alpha-\gamma+1)}{\Gamma(\alpha-\gamma+1)}\sum_{n=0}^{\infty}\frac{(\alpha)_n}{\Gamma(\gamma+n)}\frac{z^n}{n!} - \frac{1}{\Gamma(\alpha-\gamma+1)}\sum_{n=0}^{\infty}\frac{\psi(\gamma+n)\,(\alpha)_n}{\Gamma(\gamma+n)}\frac{z^n}{n!}$$

$$= \frac{1}{\Gamma(\alpha-\gamma+1)}\sum_{n=0}^{\infty}\frac{(\alpha)_n}{\Gamma(\gamma+n)}\frac{z^n}{n!}\big(\psi(\alpha-\gamma+1)-\psi(\gamma+n)\big).$$

Therefore

$$\left[\frac{\partial f}{\partial\gamma}\right]_{\gamma=s+1} = \frac{1}{\Gamma(\alpha-s)}\sum_{n=0}^{\infty}\frac{(\alpha)_n\, z^n}{n!\,(s+n)!}\big(\psi(\alpha-s)-\psi(s+n+1)\big). \qquad (8.108)$$

Similarly, by (8.103) and (8.107),

$$\frac{\partial g}{\partial\gamma} = \frac{1}{\Gamma(\alpha)}\sum_{n=0}^{\infty}\frac{z^{n+1}}{n!}\frac{\partial}{\partial\gamma}\frac{(\alpha-\gamma+1)_n\, z^{-\gamma}}{\Gamma(2-\gamma+n)} \qquad (8.109)$$

$$= \frac{1}{\Gamma(\alpha)}\sum_{n=0}^{\infty}\frac{z^{n+1-\gamma}}{n!}\left(\frac{1}{\Gamma(2-\gamma+n)}\frac{\partial}{\partial\gamma}(\alpha-\gamma+1)_n\right.$$

$$\left. - \frac{(\alpha-\gamma+1)_n}{\Gamma(2-\gamma+n)}\log z + (\alpha-\gamma+1)_n\,\frac{\psi(2-\gamma+n)}{\Gamma(2-\gamma+n)}\right).$$

In view of (8.105), we split the sum in (8.109) into the sums over the ranges $0 \leq n \leq s-1$ and $n \geq s$ (the first of these two sums being 0 if $s = 0$), and we put $\gamma = s+1$. We get

$$\left[\frac{\partial g}{\partial \gamma}\right]_{\gamma=s+1} =$$

$$\frac{1}{\Gamma(\alpha)} \sum_{n=0}^{s-1} \frac{1}{n!\, z^{s-n}} \left(\frac{1}{\Gamma(n-s+1)} \left[\frac{\partial}{\partial \gamma}(\alpha-\gamma+1)_n\right]_{\gamma=s+1} - \frac{(\alpha-s)_n}{\Gamma(n-s+1)} \log z + (\alpha-s)_n \frac{\psi(n-s+1)}{\Gamma(n-s+1)} \right)$$

$$+ \frac{1}{\Gamma(\alpha)} \sum_{n=s}^{\infty} \frac{z^{n-s}}{n!\,(n-s)!} \left(\left[\frac{\partial}{\partial \gamma}(\alpha-\gamma+1)_n\right]_{\gamma=s+1} - (\alpha-s)_n \log z + (\alpha-s)_n\, \psi(n-s+1) \right). \tag{8.110}$$

Since $\dfrac{\partial}{\partial \gamma}(\alpha-\gamma+1)_n$ is a polynomial in γ, the quantity

$$\left[\frac{\partial}{\partial \gamma}(\alpha-\gamma+1)_n\right]_{\gamma=s+1}$$

is finite. By (8.105), the only contribution to the first sum in (8.110) is given by the term

$$(\alpha-s)_n \frac{\psi(n-s+1)}{\Gamma(n-s+1)}.$$

By (6.8), for $n = 0, \ldots, s-1$ we have

$$\frac{\psi(n-s+1)}{\Gamma(n-s+1)} = -\frac{1}{\operatorname*{Res}_{t=n-s+1} \Gamma(t)} = (-1)^{s-n}(s-n-1)!.$$

Hence the first sum in (8.110) equals

$$\sum_{n=0}^{s-1} \frac{(\alpha-s)_n\,(s-n-1)!}{n!} \left(-\frac{1}{z}\right)^{s-n}.$$

To compute the second sum in (8.110) we remark that, by (6.50),

$$\begin{aligned}
\frac{\mathrm{d}}{\mathrm{d}\gamma}(\gamma)_n &= \frac{\mathrm{d}}{\mathrm{d}\gamma}\left(\gamma(\gamma+1)\cdots(\gamma+n-1)\right) \\
&= (\gamma+1)\cdots(\gamma+n-1) + \gamma(\gamma+2)\cdots(\gamma+n-1) + \ldots \\
&\qquad\qquad + \gamma(\gamma+1)\cdots(\gamma+n-2) \\
&= (\gamma)_n\left(\frac{1}{\gamma} + \frac{1}{\gamma+1} + \ldots + \frac{1}{\gamma+n-1}\right) \\
&= (\gamma)_n\left(\psi(\gamma+n) - \psi(\gamma)\right).
\end{aligned}$$

Therefore

$$\left[\frac{\partial}{\partial\gamma}(\alpha-\gamma+1)_n\right]_{\gamma=s+1} = (\alpha-s)_n\left(\psi(\alpha-s) - \psi(\alpha+n-s)\right).$$

Thus the second sum in (8.110) is

$$\begin{aligned}
&\sum_{n=s}^{\infty}\frac{(\alpha-s)_n\, z^{n-s}}{n!\,(n-s)!}\left(\psi(n-s+1) + \psi(\alpha-s) - \psi(\alpha+n-s) - \log z\right) \\
&\qquad = \sum_{n=0}^{\infty}\frac{(\alpha-s)_{s+n}\, z^n}{n!\,(s+n)!}\left(\psi(n+1) + \psi(\alpha-s) - \psi(\alpha+n) - \log z\right).
\end{aligned}$$

Hence (8.110) yields

$$\begin{aligned}
\left[\frac{\partial g}{\partial\gamma}\right]_{\gamma=s+1} &= \frac{1}{\Gamma(\alpha)}\sum_{n=0}^{s-1}\frac{(\alpha-s)_n\,(s-n-1)!}{n!}\left(-\frac{1}{z}\right)^{s-n} \qquad (8.111) \\
&+ \frac{1}{\Gamma(\alpha)}\sum_{n=0}^{\infty}\frac{(\alpha-s)_{s+n}\, z^n}{n!\,(s+n)!}\left(\psi(n+1) + \psi(\alpha-s) - \psi(\alpha+n) - \log z\right).
\end{aligned}$$

From (8.106), (8.108) and (8.111) we obtain

$$\begin{aligned}
\Psi(\alpha; s+1; z) = {} & \frac{(-1)^{s+1}}{\Gamma(\alpha-s)}\sum_{n=0}^{\infty}\frac{(\alpha)_n\, z^n}{n!\,(s+n)!}\left(\psi(\alpha-s) - \psi(s+n+1)\right) \qquad (8.112) \\
& + \frac{(-1)^s}{\Gamma(\alpha)}\sum_{n=0}^{s-1}\frac{(\alpha-s)_n\,(s-n-1)!}{n!}\left(-\frac{1}{z}\right)^{s-n} \\
& + \frac{(-1)^s}{\Gamma(\alpha)}\sum_{n=0}^{\infty}\frac{(\alpha-s)_{s+n}\, z^n}{n!\,(s+n)!} \\
& \times\left(\psi(n+1) + \psi(\alpha-s) - \psi(\alpha+n) - \log z\right).
\end{aligned}$$

Using again (8.104), from (8.112) we get the series expansion

$$\begin{aligned}\Psi(\alpha; s+1; z) = {} & \frac{1}{\Gamma(\alpha)} \sum_{n=0}^{s-1} \frac{(-1)^n (\alpha - s)_n (s-n-1)!}{n!\, z^{s-n}} \\ & + \frac{(-1)^{s+1}}{\Gamma(\alpha - s)} \sum_{n=0}^{\infty} \frac{(\alpha)_n z^n}{n!\,(s+n)!} \\ & \times \big(\psi(\alpha+n) - \psi(n+1) - \psi(s+n+1) + \log z\big) \\ & (s = 0, 1, 2, \dots;\ \alpha \neq 0, -1, -2, \dots).\end{aligned} \tag{8.113}$$

Since

$$\Phi(\alpha;\ s+1;\ z) = \sum_{n=0}^{\infty} \frac{(\alpha)_n}{(s+1)_n} \frac{z^n}{n!} = s! \sum_{n=0}^{\infty} \frac{(\alpha)_n z^n}{n!\,(s+n)!},$$

(8.113) can be written in the form

$$\Psi(\alpha;\ s+1;\ z) = \Phi(\alpha;\ s+1;\ z)\big(H \log z + \psi_2(z)\big)$$

with

$$H = \frac{(-1)^{s+1}}{s!\,\Gamma(\alpha - s)}$$

and

$$\begin{aligned}\psi_2(z) = {} & \frac{1}{\Phi(\alpha;\ s+1;\ z)} \left(\frac{1}{\Gamma(\alpha)} \sum_{n=0}^{s-1} \frac{(-1)^n (\alpha - s)_n (s-n-1)!}{n!\, z^{s-n}} \right. \\ & \left. + \frac{(-1)^{s+1}}{\Gamma(\alpha - s)} \sum_{n=0}^{\infty} \frac{(\alpha)_n z^n}{n!\,(s+n)!} \big(\psi(\alpha+n) - \psi(n+1) - \psi(s+n+1)\big) \right),\end{aligned}$$

in accordance with (7.72) and (7.73). In fact, here $s = \gamma - 1 \in \mathbb{N}$ has the same meaning as in (7.59), since in the present case the roots of the indicial equation are $r_1 = 0$ and $r_2 = 1 - \gamma$.

From (8.97) we get, for $\gamma \notin \mathbb{Z}$,

$$\Psi(\alpha; \gamma; z) = z^{1-\gamma}\, \Psi(\alpha - \gamma + 1;\ 2 - \gamma;\ z), \qquad -\pi < \arg z \leq \pi, \tag{8.114}$$

with the principal values of $\Psi(\alpha; \gamma; z)$, $\Psi(\alpha - \gamma + 1;\ 2 - \gamma;\ z)$ and $z^{1-\gamma}$.

Using (8.114), we can define $\Psi(\alpha; \gamma; z)$ for $\gamma = 0, -1, -2, \dots$. We get

$$\Psi(\alpha;\ 1 - s;\ z) = z^s\, \Psi(\alpha + s;\ s + 1;\ z), \qquad (s = 1, 2, 3, \dots). \tag{8.115}$$

By (8.97), (8.100), (8.113) and (8.115) we see that $\Psi(\alpha;\gamma;z)$ is defined for all $\alpha,\gamma\in\mathbb{C}$, as claimed.

As an analogue to Theorem 8.6, the next theorem gives an integral representation of $\Psi(\alpha;\gamma;z)$, showing that, for $\operatorname{Re}\alpha>0$, $\Psi(\alpha;\gamma;z)$ can be expressed by a (possibly oblique) Laplace transform.

Theorem 8.8 *Let* $\operatorname{Re}\alpha>0$, *and let* $z\in\mathbb{C}\backslash\{0\}$. *Let* $\arg z$ *denote the principal argument of* z, *i.e.*, $-\pi<\arg z\le\pi$. *For any* φ *satisfying*

$$-\arg z-\frac{\pi}{2}<\varphi<-\arg z+\frac{\pi}{2}\quad\textit{and}\quad -\pi<\varphi<\pi, \tag{8.116}$$

the following integral representation holds:

$$\Psi(\alpha;\gamma;z)=\frac{1}{\Gamma(\alpha)}\int_0^{e^{i\varphi}\infty}e^{-zt}\,t^{\alpha-1}(1+t)^{\gamma-\alpha-1}\,\mathrm{d}t, \tag{8.117}$$

where the path of integration is the half-line from zero to infinity through $e^{i\varphi}$. *Accordingly,*

$$t^{\alpha-1}=\exp\big((\alpha-1)(\log|t|+i\arg t)\big)$$

with $\arg t=\varphi$, *and similarly*

$$(1+t)^{\gamma-\alpha-1}=\exp\big((\gamma-\alpha-1)(\log|1+t|+i\arg(1+t))\big)$$

with $\arg 1=0$ *and* $\arg(1+t)\to\varphi$ *as* $t\to\infty$.

If $\operatorname{Re}z>0$, *i.e., if* $-\pi/2<\arg z<\pi/2$, *one can choose* $\varphi=0$ *in* (8.116), *so that the integral in* (8.117) *can be taken to be the ordinary Laplace transform of* $t^{\alpha-1}(1+t)^{\gamma-\alpha-1}$.

Proof By (8.116), $\cos(\arg z+\varphi)>0$. Since

$$\left|e^{-zt}\right|=e^{-|zt|\cos(\arg z+\varphi)}$$

and

$$\left|t^{\alpha-1}\right|=e^{-\varphi\operatorname{Im}\alpha}\,|t|^{\operatorname{Re}\alpha-1},$$

by the assumption $\operatorname{Re}\alpha>0$ the integral in (8.117) is absolutely and uniformly convergent for z in a neighbourhood of any point z_0 satisfying $-\pi/2<\arg z_0+\varphi<\pi/2$. Therefore, denoting the right-hand side of (8.117) by

$$w:=\frac{1}{\Gamma(\alpha)}\int_0^{e^{i\varphi}\infty}e^{-zt}\,t^{\alpha-1}(1+t)^{\gamma-\alpha-1}\,\mathrm{d}t, \tag{8.118}$$

we get

$$\frac{\mathrm{d}w}{\mathrm{d}z} = -\frac{1}{\Gamma(\alpha)} \int_0^{e^{i\varphi}\infty} e^{-zt}\, t^{\alpha}(1+t)^{\gamma-\alpha-1}\,\mathrm{d}t \tag{8.119}$$

and

$$\frac{\mathrm{d}^2 w}{\mathrm{d}z^2} = \frac{1}{\Gamma(\alpha)} \int_0^{e^{i\varphi}\infty} e^{-zt}\, t^{\alpha+1}(1+t)^{\gamma-\alpha-1}\,\mathrm{d}t,$$

whence

$$\begin{aligned} z\,\frac{\mathrm{d}^2 w}{\mathrm{d}z^2} &+ (\gamma - z)\frac{\mathrm{d}w}{\mathrm{d}z} - \alpha w \\ &= \frac{1}{\Gamma(\alpha)} \int_0^{e^{i\varphi}\infty} e^{-zt}\, t^{\alpha-1}(1+t)^{\gamma-\alpha-1}\big(zt^2 - (\gamma - z)t - \alpha\big)\,\mathrm{d}t \\ &= -\frac{1}{\Gamma(\alpha)} \int_0^{e^{i\varphi}\infty} \frac{\mathrm{d}}{\mathrm{d}t}\big(e^{-zt}\, t^{\alpha}(1+t)^{\gamma-\alpha}\big)\,\mathrm{d}t \\ &= -\frac{1}{\Gamma(\alpha)} \Big[e^{-zt}\, t^{\alpha}(1+t)^{\gamma-\alpha} \Big]_{t=0}^{t=e^{i\varphi}\infty} = 0. \end{aligned}$$

Thus the function w defined in (8.118) is a solution of the confluent hypergeometric equation (8.86). By (8.94), if $\gamma \notin \mathbb{Z}$ we get

$$w = C_1\, \Phi(\alpha; \gamma; z) + C_2\, z^{1-\gamma}\, \Phi(\alpha - \gamma + 1;\ 2 - \gamma;\ z), \tag{8.120}$$

where $C_1 = C_1(\alpha, \gamma)$ and $C_2 = C_2(\alpha, \gamma)$ are independent of z.

If $\operatorname{Re}\gamma < 1$ we have

$$\left|z^{1-\gamma}\right| = e^{(\arg z)\operatorname{Im}\gamma}\, |z|^{1-\operatorname{Re}\gamma} \le e^{\pi \operatorname{Im}\gamma}\, |z|^{1-\operatorname{Re}\gamma},$$

whence

$$\lim_{z\to 0} \left|z^{1-\gamma}\right| = 0.$$

We temporarily assume $0 < \operatorname{Re}\gamma < 1$. From (8.118) and (8.120) we obtain

$$C_1 = \lim_{z\to 0} w = \frac{1}{\Gamma(\alpha)} \int_0^{e^{i\varphi}\infty} t^{\alpha-1}(1+t)^{\gamma-\alpha-1}\,\mathrm{d}t, \tag{8.121}$$

where the interchange of limit and integral in (8.118) is allowed by uniform convergence of (8.118) near $z = 0$, because $\left|e^{-zt}\right| = e^{-|zt|\cos(\arg z+\varphi)} < 1$, and

$$\left|t^{\alpha-1}(1+t)^{\gamma-\alpha-1}\right| \ll \begin{cases} |\,t\,|^{\operatorname{Re}\alpha-1} & \text{for } t \to 0 \\ |\,t\,|^{\operatorname{Re}\gamma-2} & \text{for } t \to \infty. \end{cases}$$

For any real number $r > 0$, let λ_r denote the arc

$$\lambda_r := \{t \in \mathbb{C} \mid |\,t\,| = r, \ \ \arg t \ \text{from 0 to } \varphi\}.$$

Since

$$\left| \int\limits_{\lambda_r} t^{\alpha-1}(1+t)^{\gamma-\alpha-1}\,\mathrm{d}t \right| \le \int\limits_{\lambda_r} \left|t^{\alpha-1}\right| \left|(1+t)^{\gamma-\alpha-1}\right| |\mathrm{d}t|$$

$$\ll r^{\operatorname{Re}\alpha-1}(1+r)^{\operatorname{Re}\gamma-\operatorname{Re}\alpha-1}\,2\pi r \ll \begin{cases} r^{\operatorname{Re}\alpha} & \text{for } r \to 0 \\ r^{\operatorname{Re}\gamma-1} & \text{for } r \to +\infty, \end{cases}$$

our assumptions $\operatorname{Re}\alpha > 0$ and $\operatorname{Re}\gamma < 1$ yield

$$\lim_{\varepsilon\to 0} \int\limits_{\lambda_\varepsilon} t^{\alpha-1}(1+t)^{\gamma-\alpha-1}\,\mathrm{d}t = \lim_{R\to+\infty} \int\limits_{\lambda_R} t^{\alpha-1}(1+t)^{\gamma-\alpha-1}\,\mathrm{d}t = 0. \tag{8.122}$$

By Cauchy's theorem,

$$\left(\int\limits_{\varepsilon}^{R} + \int\limits_{\lambda_R} - \int\limits_{\varepsilon e^{i\varphi}}^{Re^{i\varphi}} - \int\limits_{\lambda_\varepsilon} \right) t^{\alpha-1}(1+t)^{\gamma-\alpha-1}\,\mathrm{d}t = 0.$$

Hence, by (8.122),

$$\int\limits_{0}^{e^{i\varphi}\infty} t^{\alpha-1}(1+t)^{\gamma-\alpha-1}\,\mathrm{d}t = \int\limits_{0}^{+\infty} t^{\alpha-1}(1+t)^{\gamma-\alpha-1}\,\mathrm{d}t. \tag{8.123}$$

From (6.11) and (6.13) we get

$$\int\limits_{0}^{+\infty} t^{\alpha-1}(1+t)^{\gamma-\alpha-1}\,\mathrm{d}t = \frac{\Gamma(\alpha)\Gamma(1-\gamma)}{\Gamma(\alpha-\gamma+1)}. \tag{8.124}$$

Therefore, by (8.121), (8.123) and (8.124),

$$C_1 = \frac{\Gamma(1-\gamma)}{\Gamma(\alpha-\gamma+1)}. \tag{8.125}$$

From (8.120) and (8.80) we have

$$\begin{aligned} z^\gamma \frac{\mathrm{d}w}{\mathrm{d}z} &= C_1 \frac{\alpha}{\gamma} z^\gamma \Phi(\alpha+1;\ \gamma+1;\ z) + C_2(1-\gamma)\,\Phi(\alpha-\gamma+1;\ 2-\gamma;\ z) \\ &\quad + C_2 \frac{\alpha-\gamma+1}{2-\gamma} z\,\Phi(\alpha-\gamma+2;\ 3-\gamma;\ z). \end{aligned}$$

By the assumption $\operatorname{Re}\gamma > 0$ and by (8.119) we get

$$C_2(1-\gamma) = -\frac{1}{\Gamma(\alpha)} \lim_{z\to 0} z^\gamma \int_0^{e^{i\varphi}\infty} e^{-zt}\, t^\alpha (1+t)^{\gamma-\alpha-1}\, \mathrm{d}t.$$

With the substitution $zt = \tau$ we obtain

$$z^\gamma \int_0^{e^{i\varphi}\infty} e^{-zt}\, t^\alpha (1+t)^{\gamma-\alpha-1}\, \mathrm{d}t = \int_0^{e^{i(\arg z+\varphi)}\infty} e^{-\tau}\tau^\alpha (z+\tau)^{\gamma-\alpha-1}\, \mathrm{d}\tau.$$

As above, we can plainly rotate the integration path for τ from the half-line $\left(0,\ e^{i(\arg z+\varphi)}\infty\right)$ to $(0, +\infty)$. Hence, by uniform convergence near $z = 0$,

$$\begin{aligned} C_2 &= \frac{1}{(\gamma-1)\Gamma(\alpha)} \lim_{z\to 0} \int_0^{+\infty} e^{-\tau}\tau^\alpha (z+\tau)^{\gamma-\alpha-1}\, \mathrm{d}\tau \\ &= \frac{1}{(\gamma-1)\Gamma(\alpha)} \int_0^{+\infty} e^{-\tau}\tau^{\gamma-1}\, \mathrm{d}\tau = \frac{\Gamma(\gamma)}{(\gamma-1)\Gamma(\alpha)} = \frac{\Gamma(\gamma-1)}{\Gamma(\alpha)}. \end{aligned} \tag{8.126}$$

Thus (8.97), (8.118), (8.120), (8.125) and (8.126) yield, for $\operatorname{Re}\alpha > 0$ and $0 < \operatorname{Re}\gamma < 1$,

$$\begin{aligned} &\frac{1}{\Gamma(\alpha)} \int_0^{e^{i\varphi}\infty} e^{-zt}\, t^{\alpha-1}(1+t)^{\gamma-\alpha-1}\, \mathrm{d}t \\ &= \frac{\Gamma(1-\gamma)}{\Gamma(\alpha-\gamma+1)} \Phi(\alpha;\gamma;z) + \frac{\Gamma(\gamma-1)}{\Gamma(\alpha)} z^{1-\gamma}\,\Phi(\alpha-\gamma+1;\ 2-\gamma;\ z) \\ &= \Psi(\alpha;\gamma;z), \end{aligned}$$

i.e., (8.117). Clearly, by analytic continuation on γ, the temporary assumption $0 < \operatorname{Re}\gamma < 1$ can be dropped. This completes the proof of the theorem. □

With the next corollary we prove for the function $\Psi(\alpha;\gamma;z)$ an analogue of the differentiation formula (8.80) for $\Phi(\alpha;\gamma;z)$.

Corollary 8.1

$$\frac{\mathrm{d}^m}{\mathrm{d}z^m}\Psi(\alpha;\gamma;z) = (-1)^m(\alpha)_m\Psi(\alpha+m;\ \gamma+m;\ z)\quad (m=1,2,3,\ldots). \tag{8.127}$$

Proof If $\operatorname{Re}\alpha > 0$, by (8.117), (8.118) and (8.119) we get

$$\begin{aligned}\frac{\mathrm{d}}{\mathrm{d}z}\Psi(\alpha;\gamma;z) &= -\frac{1}{\Gamma(\alpha)}\int_0^{e^{i\varphi}\infty} e^{-zt}\,t^{\alpha}(1+t)^{\gamma-\alpha-1}\,\mathrm{d}t\\ &= -\frac{\alpha}{\Gamma(\alpha+1)}\int_0^{e^{i\varphi}\infty} e^{-zt}\,t^{\alpha}(1+t)^{\gamma-\alpha-1}\,\mathrm{d}t\\ &= -\alpha\,\Psi(\alpha+1;\ \gamma+1;\ z).\end{aligned}$$

By analytic continuation on α, the differentiation formula

$$\frac{\mathrm{d}}{\mathrm{d}z}\Psi(\alpha;\gamma;z) = -\alpha\,\Psi(\alpha+1;\ \gamma+1;\ z) \tag{8.128}$$

holds for all $\alpha,\gamma\in\mathbb{C}$. Then (8.127) follows from (8.128) by induction on m. □

Using (8.96), (8.98) and (8.128) we can compute the wronskian of $\Phi^*(\alpha;\gamma;z)$ and $\Psi(\alpha;\gamma;z)$. A straightforward computation yields

$$\begin{aligned}&W\big(\Phi^*(\alpha;\gamma;z),\ \Psi(\alpha;\gamma;z)\big)\\ &\quad= -\frac{\pi z^{-\gamma}}{\Gamma(\alpha)\sin(\pi\gamma)}\big(\Phi^*(\alpha;\gamma;z)\,\Phi^*(\alpha-\gamma+1;\ 1-\gamma;\ z)\\ &\qquad\qquad -\alpha z\,\Phi^*(\alpha+1;\ \gamma+1;\ z)\,\Phi^*(\alpha-\gamma+1;\ 2-\gamma;\ z)\big)\\ &\quad= Ce^{z}z^{-\gamma}\end{aligned}$$

by (8.95). Multiplying by z^{γ} and then putting $z=0$ we get

$$\begin{aligned}C &= -\frac{\pi}{\Gamma(\alpha)\sin(\pi\gamma)}\Phi^*(\alpha;\gamma;0)\,\Phi^*(\alpha-\gamma+1;\ 1-\gamma;\ 0)\\ &= -\frac{\pi}{\Gamma(\alpha)\sin(\pi\gamma)}\frac{1}{\Gamma(\gamma)}\frac{1}{\Gamma(1-\gamma)} = -\frac{1}{\Gamma(\alpha)}.\end{aligned}$$

Therefore

$$W\big(\Phi^*(\alpha;\gamma;z),\ \Psi(\alpha;\gamma;z)\big) = -\frac{1}{\Gamma(\alpha)}\, e^z z^{-\gamma}.$$

This shows that $\Phi^*(\alpha;\gamma;z)$ and $\Psi(\alpha;\gamma;z)$ are linearly independent if and only if $\alpha \neq 0, -1, -2, \ldots$, as claimed.

8.8 Mellin–Barnes' Integral Representations

Let $f : \mathbb{R}^+ \to \mathbb{C}$ and $\sigma \in \mathbb{R}$ satisfy

$$\int_0^{+\infty} |f(x)|\, x^{\sigma-1}\, \mathrm{d}x \ < +\infty, \tag{8.129}$$

and let

$$(\sigma) = \{s \in \mathbb{C} \,|\, s = \sigma + it,\ -\infty < t < +\infty\}$$

denote the vertical line in $\mathbb{C}$ of real part σ.

The Mellin transform of f is the function $F : (\sigma) \to \mathbb{C}$ defined by

$$F(s) = \int_0^{+\infty} f(x)\, x^{s-1}\, \mathrm{d}x, \qquad s = \sigma + it. \tag{8.130}$$

By (8.129), the integral (8.130) is absolutely convergent.

With the substitution $x = e^{-2\pi u}$ the integral (8.130) becomes

$$F(\sigma + it) = 2\pi \int_{-\infty}^{+\infty} f\big(e^{-2\pi u}\big) e^{-2\pi(\sigma+it)u}\, \mathrm{d}u = 2\pi\, \widehat{\varphi_\sigma}(t), \tag{8.131}$$

where

$$\widehat{\varphi_\sigma}(t) = \int_{-\infty}^{+\infty} \varphi_\sigma(u) e^{-2\pi i t u}\, \mathrm{d}u$$

is the Fourier transform of the function φ_σ defined by

$$\varphi_\sigma(u) = f\big(e^{-2\pi u}\big) e^{-2\pi\sigma u} \quad \text{for } -\infty < u < +\infty.$$

Therefore, under mild assumptions for f besides (8.129), applying Fourier's integral theorem $\widehat{\widehat{\varphi_\sigma}}(u) = \varphi_\sigma(-u)$ we get

$$f\left(e^{-2\pi u}\right)e^{-2\pi\sigma u} = \varphi_\sigma(u) = \widehat{\widehat{\varphi_\sigma}}(-u)$$
$$= \int_{-\infty}^{+\infty} \widehat{\varphi_\sigma}(t)e^{2\pi iut}\,\mathrm{d}t = \frac{1}{2\pi}\int_{-\infty}^{+\infty} F(\sigma+it)e^{2\pi iut}\,\mathrm{d}t,$$

whence

$$f\left(e^{-2\pi u}\right) = \frac{1}{2\pi i}\int_{t=-\infty}^{+\infty} F(\sigma+it)e^{2\pi u(\sigma+it)}\mathrm{d}(\sigma+it),$$

i.e., the Mellin inversion formula:

$$f(x) = \frac{1}{2\pi i}\int_{(\sigma)} F(s)\,x^{-s}\,\mathrm{d}s, \qquad x > 0. \tag{8.132}$$

A standard example of Mellin's transform is the gamma-function. Choosing $f(x) = e^{-x}$ in (8.130), for $\sigma = \operatorname{Re} s > 0$ we get $F(s) = \Gamma(s)$ from (6.1), so that $\Gamma(s)$ is the Mellin transform of e^{-x}. Thus the Mellin inversion formula (8.132) yields

$$e^{-x} = \frac{1}{2\pi i}\int_{\sigma-i\infty}^{\sigma+i\infty} \Gamma(s)\,x^{-s}\,\mathrm{d}s \qquad (x > 0,\ \sigma > 0).$$

By analytic continuation we obtain

$$e^{-z} = \frac{1}{2\pi i}\int_{\sigma-i\infty}^{\sigma+i\infty} \Gamma(s)\,z^{-s}\,\mathrm{d}s \qquad (z \in \mathbb{C},\ \operatorname{Re} z > 0,\ \sigma > 0). \tag{8.133}$$

Some formulae, similar to (8.133), representing hypergeometric functions as Mellin integrals over vertical lines in $\mathbb{C}$ with integrands containing gamma-factors were first proved by Barnes through a method, independent of Fourier analysis, based on the residue theorem and the exponential decay of $|\Gamma(s)|$ along vertical strips expressed by (6.42).

We follow Barnes' method, beginning with the Kummer confluent hypergeometric function $\Phi(\alpha;\gamma;z) = {}_1F_1(\alpha;\gamma;z)$.

Theorem 8.9 *Let $k \in \mathbb{R}$, and let $\alpha, \gamma, z \in \mathbb{C}$ with $\alpha, \gamma \neq 0, -1, -2, \dots$ and* $\operatorname{Re} z < 0$. *Then*

$$\Phi(\alpha;\gamma;z) = {}_1F_1(\alpha;\gamma;z) = \frac{\Gamma(\gamma)}{\Gamma(\alpha)} \frac{1}{2\pi i} \int_{(k)} \frac{\Gamma(s+\alpha)}{\Gamma(s+\gamma)} \Gamma(-s)(-z)^s \, ds, \quad (8.134)$$

where (k) denotes the vertical line $(k - i\infty,\ k + i\infty)$ deformed (if necessary) inside a bounded region to separate the poles $s = 0, 1, 2, \dots$ of $\Gamma(-s)$ from the poles $s = -\alpha,\ -\alpha - 1,\ -\alpha - 2, \dots$ of $\Gamma(s+\alpha)$, so that $s = 0, 1, 2, \dots$ lie on the right and $s = -\alpha,\ -\alpha - 1,\ -\alpha - 2, \dots$ lie on the left of the integration path (this is obviously possible for any $k \in \mathbb{R}$, since $\alpha \neq 0, -1, -2, \dots$). In (8.134), $(-z)^s =$ $\exp\big(s \log(-z)\big)$ *with* $\log(-z) \in \mathbb{R}$ *for* $z \in \mathbb{R}$, $z < 0$.

Proof Let $-z = \varrho e^{i\vartheta}$, $\alpha = \alpha_1 + i\alpha_2$, $\gamma = \gamma_1 + i\gamma_2$ and $s = k + it$ for sufficiently large $|t|$. Then

$$|(-z)^s| = \exp \operatorname{Re}\big(s \log(-z)\big) = \exp \operatorname{Re}\big((k+it)(\log \varrho + i\vartheta)\big) = \varrho^k e^{-\vartheta t}. \quad (8.135)$$

Since k is fixed, for bounded $\varrho = |z|$ we get by (6.42), for any $\varepsilon > 0$ and for $t \to \pm\infty$,

$$\begin{aligned} &\left| \frac{\Gamma(s+\alpha)}{\Gamma(s+\gamma)} \Gamma(-s)(-z)^s \right| \qquad (8.136) \\ &\quad \ll |t|^{\alpha_1 - \gamma_1 - k - \frac{1}{2}} \exp\Big(-\frac{\pi}{2}\big(|t+\alpha_2| - |t+\gamma_2| + |t|\big) - \vartheta t\Big) \\ &\quad \ll |t|^{\alpha_1 - \gamma_1 - k - \frac{1}{2}} \exp\Big(\Big(|\vartheta| - \frac{\pi}{2} + \varepsilon\Big)|t|\Big). \end{aligned}$$

Since $\operatorname{Re} z < 0$ we have $|\vartheta| < \pi/2$. We take ε such that $0 < \varepsilon < \pi/2 - |\vartheta|$. Thus, by (8.136), the integral (8.134) is absolutely and uniformly convergent for z in any bounded region contained in the half-plane $\operatorname{Re} z < 0$, and hence is a regular function of z in $\operatorname{Re} z < 0$.

For an integer $N > \max\{k, |\alpha_1|\}$ and for a real $T > |\alpha_2|$ we consider the integral, with the same integrand as in (8.134), taken along the border of the region $\mathcal{R}$ in the s-plane obtained from the rectangle with vertices at $k \pm iT$ and $N + 1/2 \pm iT$ by deforming (if necessary) the left side $(k - iT,\ k + iT)$ to exclude from $\mathcal{R}$ the poles of $\Gamma(s+\alpha)$ and to include in $\mathcal{R}$ the poles $s = 0, 1, \dots, N$ of $\Gamma(-s)$, as in the statement of the theorem. If we keep N fixed and let $T \to +\infty$, from (8.136) we see that the integrals along the upper and lower sides $(k + iT, N + 1/2 + iT)$ and $(k - iT,\ N + 1/2 - iT)$ of $\mathcal{R}$ tend to zero. Therefore, by the residue theorem, the difference

$$\frac{1}{2\pi i}\Bigg(\int_{N+\frac{1}{2}-i\infty}^{N+\frac{1}{2}+i\infty} - \int_{(k)} \Bigg) \frac{\Gamma(s+\alpha)}{\Gamma(s+\gamma)} \Gamma(-s)(-z)^s \, ds \quad (8.137)$$

equals the sum of residues of the integrand at the poles $s = 0, 1, \ldots, N$ of $\Gamma(-s)$. By (6.8), the residue of $\Gamma(-s)$ at $s = n$ $(0 \le n \le N)$ is $(-1)^{n+1}/n!$. Thus (8.137) equals, by (8.17),

$$\sum_{n=0}^{N} \frac{(-1)^{n+1}}{n!} \frac{\Gamma(\alpha+n)}{\Gamma(\gamma+n)}(-z)^n \tag{8.138}$$

$$= -\sum_{n=0}^{N} \frac{(\alpha)_n\, \Gamma(\alpha)}{(\gamma)_n\, \Gamma(\gamma)} \frac{z^n}{n!} = -\frac{\Gamma(\alpha)}{\Gamma(\gamma)} \sum_{n=0}^{N} \frac{(\alpha)_n}{(\gamma)_n} \frac{z^n}{n!}.$$

We now show that the integral over $(N+1/2-i\infty,\ N+1/2+i\infty)$ in (8.137) tends to zero as $N \to +\infty$. Let

$$I_N = \frac{1}{2\pi i} \int_{N+\frac{1}{2}-i\infty}^{N+\frac{1}{2}+i\infty} \frac{\Gamma(s+\alpha)}{\Gamma(s+\gamma)} \Gamma(-s)(-z)^s \,\mathrm{d}s.$$

With the substitution $s = N + 1/2 + it$, using (8.135), we get

$$|I_N| \ll \int_{-\infty}^{+\infty} \left| \frac{\Gamma(\alpha + N + 1/2 + it)}{\Gamma(\gamma + N + 1/2 + it)} \right| |\Gamma(-N - 1/2 - it)|\, |z|^{N+\frac{1}{2}} e^{-\vartheta t} \,\mathrm{d}t.$$

By Euler's reflection formula (6.21), since $|\sin(\pi(N+1/2+it))| = \cosh(\pi t)$, we obtain

$$|\Gamma(-N-1/2-it)| = \frac{\pi}{\cosh(\pi t)\, |\Gamma(N+3/2+it)|}.$$

Hence

$$|I_N| \ll |z|^{N+\frac{1}{2}} \tag{8.139}$$

$$\times \int_{-\infty}^{+\infty} \left| \frac{\Gamma(\alpha + N + 1/2 + it)}{\Gamma(\gamma + N + 1/2 + it)} \right| \frac{1}{|\Gamma(N+3/2+it)|} \frac{e^{-\vartheta t}}{\cosh(\pi t)} \,\mathrm{d}t.$$

By repeated application of the functional equation (6.6) we get

$$\Gamma\left(N + \frac{3}{2} + it\right) = \left(\frac{1}{2} + it\right)\left(\frac{3}{2} + it\right) \cdots \left(N + \frac{1}{2} + it\right) \Gamma\left(\frac{1}{2} + it\right),$$

whence, by (6.23),

$$\begin{aligned}\left|\Gamma\Big(N+\frac{3}{2}+it\Big)\right| &\geq \frac{1}{2}\cdot\frac{3}{2}\cdots\frac{2N+1}{2}\left|\Gamma\Big(\frac{1}{2}+it\Big)\right| \\ &= \frac{(2N+1)!!}{2^{N+1}}\left|\Gamma\Big(\frac{1}{2}+it\Big)\right| = \frac{1}{\sqrt{\pi}}\,\Gamma\Big(N+\frac{3}{2}\Big)\left|\Gamma\Big(\frac{1}{2}+it\Big)\right|.\end{aligned} \tag{8.140}$$

Again by the reflection formula (6.21) we get

$$\left|\Gamma\Big(\frac{1}{2}+it\Big)\right|^2 = \Gamma\Big(\frac{1}{2}+it\Big)\Gamma\Big(\frac{1}{2}-it\Big) = \frac{\pi}{\sin(\pi/2+i\pi t)} = \frac{\pi}{\cosh(\pi t)}.$$

Thus (8.140) yields

$$\left|\Gamma\Big(N+\frac{3}{2}+it\Big)\right| \geq \frac{\Gamma(N+3/2)}{\sqrt{\cosh(\pi t)}}. \tag{8.141}$$

By Stirling's formula (6.40) with z, α replaced by $N+it$ and $\alpha+1/2$ respectively we obtain

$$\begin{aligned}&\log\Gamma(\alpha+N+1/2+it)\\ &\quad = (\alpha+N+it)\log(N+it) - N - it + \log\sqrt{2\pi} + O\Big(\frac{1}{|N+it|}\Big),\end{aligned}$$

whence

$$\Gamma(\alpha+N+1/2+it) = \sqrt{2\pi}\,(N+it)^{\alpha+N+it}e^{-N-it}\Big(1+O\Big(\frac{1}{|N+it|}\Big)\Big), \tag{8.142}$$

and similarly for $\Gamma(\gamma+N+1/2+it)$. Therefore

$$\left|\frac{\Gamma(\alpha+N+1/2+it)}{\Gamma(\gamma+N+1/2+it)}\right| = \left|(N+it)^{\alpha-\gamma}\right|\left|1+O\Big(\frac{1}{|N+it|}\Big)\right|.$$

As with (8.135) we have

$$\begin{aligned}\left|(N+it)^{\alpha-\gamma}\right| &= \left|N^{\alpha-\gamma}\right|\left|\Big(1+\frac{it}{N}\Big)^{\alpha-\gamma}\right| \\ &= N^{\alpha_1-\gamma_1}\left|1+\frac{it}{N}\right|^{\alpha_1-\gamma_1}\exp\Big(-\arg\Big(1+\frac{it}{N}\Big)(\alpha_2-\gamma_2)\Big) \\ &\leq N^{\alpha_1-\gamma_1}\left|1+\frac{it}{N}\right|^{\alpha_1-\gamma_1}\exp\Big(\frac{\pi}{2}\,|\alpha_2-\gamma_2|\Big).\end{aligned}$$

Thus, for $N \to +\infty$,

$$\left|\frac{\Gamma(\alpha+N+1/2+it)}{\Gamma(\gamma+N+1/2+it)}\right| \ll N^{\alpha_1-\gamma_1}\left|1+\frac{it}{N}\right|^{\alpha_1-\gamma_1} \tag{8.143}$$
$$\leq N^{\alpha_1-\gamma_1}\,\lambda_{\alpha_1-\gamma_1}(t),$$

where

$$\lambda_\nu(t) = \begin{cases} \left(1+t^2\right)^{\nu/2} & \text{for } \nu > 0 \\ 1 & \text{for } \nu \leq 0. \end{cases} \tag{8.144}$$

From (8.139), (8.141) and (8.143) we get

$$|I_N| \ll \frac{|z|^{N+\frac{1}{2}} N^{\alpha_1-\gamma_1}}{\Gamma(N+3/2)} \int_{-\infty}^{+\infty} \lambda_{\alpha_1-\gamma_1}(t)\,\frac{e^{-\vartheta t}}{\sqrt{\cosh(\pi t)}}\,\mathrm{d}t. \tag{8.145}$$

Since $\sqrt{\cosh(\pi t)} \sim e^{\frac{\pi}{2}|t|}/\sqrt{2}$ as $t \to \pm\infty$, the last integral converges for $|\vartheta| = |\arg(-z)| < \pi/2$, i.e., for $\operatorname{Re} z < 0$, which we have assumed. Moreover, again by Stirling's formula (6.40), $\Gamma(N+3/2) \gg N^{N+1}e^{-N}$. Thus (8.145) yields

$$\lim_{N\to+\infty} I_N = 0$$

as claimed. Then, making $N \to +\infty$ in (8.137), (8.138), we get

$$-\frac{1}{2\pi i}\int_{(k)} \frac{\Gamma(s+\alpha)}{\Gamma(s+\gamma)}\,\Gamma(-s)(-z)^s\,\mathrm{d}s = -\frac{\Gamma(\alpha)}{\Gamma(\gamma)}\sum_{n=0}^{\infty}\frac{(\alpha)_n}{(\gamma)_n}\frac{z^n}{n!},$$

i.e., (8.134). □

We remark that, by substituting $s \mapsto -s$, $z \mapsto -z$ and choosing $\alpha = \gamma$ in (8.134), we obtain (8.133).

Next we prove Barnes' integral representation of ${}_2F_1(\alpha, \beta; \gamma; z)$. The argument is similar to the proof of Theorem 8.9 except that for the proof of Theorem 8.10 we shall require the temporary assumption $|z| < 1$, which ensures convergence of the hypergeometric series (8.12).

Theorem 8.10 *Let $k \in \mathbb{R}$, let $\alpha, \beta, \gamma \in \mathbb{C}\setminus\{0, -1, -2, \ldots\}$ and $z \in \mathbb{C}\setminus[0, +\infty)$. Then*

$${}_2F_1(\alpha, \beta; \gamma; z) = \frac{\Gamma(\gamma)}{\Gamma(\alpha)\Gamma(\beta)}\frac{1}{2\pi i}\int_{(k)} \frac{\Gamma(s+\alpha)\Gamma(s+\beta)}{\Gamma(s+\gamma)}\,\Gamma(-s)(-z)^s\,\mathrm{d}s, \tag{8.146}$$

where (k) *denotes the vertical line* $(k - i\infty, k + i\infty)$ *deformed (if necessary) inside a bounded region to separate the poles of* $\Gamma(-s)$ *from the poles of* $\Gamma(s+\alpha)\Gamma(s+\beta)$.

Proof Let $-z = \varrho e^{i\vartheta}$, $\alpha = \alpha_1 + i\alpha_2$, $\beta = \beta_1 + i\beta_2$, $\gamma = \gamma_1 + i\gamma_2$ and $s = k + it$ for sufficiently large $|t|$. By (6.42) and (8.135) we get, for bounded $\varrho = |z|$, for any $\varepsilon > 0$ and for $t \to \pm\infty$,

$$\left|\frac{\Gamma(s+\alpha)\Gamma(s+\beta)}{\Gamma(s+\gamma)}\Gamma(-s)(-z)^s\right| \tag{8.147}$$
$$\ll |t|^{\alpha_1+\beta_1-\gamma_1-1}\exp\left(-\frac{\pi}{2}(|t+\alpha_2| + |t+\beta_2| - |t+\gamma_2| + |t|) - \vartheta t\right)$$
$$\ll |t|^{\alpha_1+\beta_1-\gamma_1-1}\exp\left((|\vartheta| - \pi + \varepsilon)|t|\right).$$

Since $z \in \mathbb{C}\backslash[0, +\infty)$ we have $|\vartheta| = |\arg(-z)| < \pi$. We take ε such that $0 < \varepsilon < \pi - |\vartheta|$. By (8.147) the integral on the right-hand side of (8.146) is a regular function of z in $\mathbb{C}\backslash[0, +\infty)$.

For an integer $N > \max\{k, |\alpha_1|, |\beta_1|\}$ and for a real $T > \max\{|\alpha_2|, |\beta_2|\}$, let $\mathcal{R}$ be the region in the s-plane obtained from the rectangle with vertices at $k \pm iT$ and $N + 1/2 \pm iT$ by deforming (if necessary) the left side $(k - iT, k + iT)$ to exclude from $\mathcal{R}$ the poles of $\Gamma(s+\alpha)\Gamma(s+\beta)$ and to include in $\mathcal{R}$ the poles $s = 0, 1, \dots, N$ of $\Gamma(-s)$. For fixed N and for $T \to +\infty$, by (8.147) the integrals along the upper and lower sides of $\mathcal{R}$ tend to zero. Thus the difference

$$\frac{1}{2\pi i}\left(\int_{N+\frac{1}{2}-i\infty}^{N+\frac{1}{2}+i\infty} - \int_{(k)}\right)\frac{\Gamma(s+\alpha)\Gamma(s+\beta)}{\Gamma(s+\gamma)}\Gamma(-s)(-z)^s\,\mathrm{d}s \tag{8.148}$$

equals the sum of residues of the integrand at the poles $s = 0, 1, \dots, N$ of $\Gamma(-s)$. Hence (8.148) equals

$$\sum_{n=0}^{N}\frac{(-1)^{n+1}}{n!}\frac{\Gamma(\alpha+n)\Gamma(\beta+n)}{\Gamma(\gamma+n)}(-z)^n \tag{8.149}$$
$$= -\frac{\Gamma(\alpha)\Gamma(\beta)}{\Gamma(\gamma)}\sum_{n=0}^{N}\frac{(\alpha)_n(\beta)_n}{(\gamma)_n}\frac{z^n}{n!}.$$

Let

$$J_N = \frac{1}{2\pi i}\int_{N+\frac{1}{2}-i\infty}^{N+\frac{1}{2}+i\infty}\frac{\Gamma(s+\alpha)\Gamma(s+\beta)}{\Gamma(s+\gamma)}\Gamma(-s)(-z)^s\,\mathrm{d}s.$$

We substitute $s = N + 1/2 + it$. Similarly to (8.139) we obtain

$$|J_N| \ll |z|^{N+\frac{1}{2}} \times \int_{-\infty}^{+\infty} \left| \frac{\Gamma(\alpha+N+1/2+it)\,\Gamma(\beta+N+1/2+it)}{\Gamma(\gamma+N+1/2+it)\,\Gamma(N+3/2+it)} \right| \frac{e^{-\vartheta t}}{\cosh(\pi t)}\,dt. \tag{8.150}$$

By Stirling's formula (6.40) with z replaced by $N + it$ and α replaced by $3/2$ we get

$$\Gamma(N+3/2+it) = \sqrt{2\pi}\,(N+it)^{N+1+it} e^{-N-it}\left(1 + O\left(\frac{1}{|N+it|}\right)\right).$$

Combining this with (8.142) and the analogous asymptotic formulae with β and γ in place of α we obtain

$$\begin{aligned}
&\left| \frac{\Gamma(\alpha+N+1/2+it)\,\Gamma(\beta+N+1/2+it)}{\Gamma(\gamma+N+1/2+it)\,\Gamma(N+3/2+it)} \right| \ll \left|(N+it)^{\alpha+\beta-\gamma-1}\right| \\
&= |N^{\alpha+\beta-\gamma-1}| \left| \left(1+\frac{it}{N}\right)^{\alpha+\beta-\gamma-1} \right| \\
&= N^{\alpha_1+\beta_1-\gamma_1-1} \left|1+\frac{it}{N}\right|^{\alpha_1+\beta_1-\gamma_1-1} \exp\left(-\arg\left(1+\frac{it}{N}\right)(\alpha_2+\beta_2-\gamma_2)\right) \\
&\le N^{\alpha_1+\beta_1-\gamma_1-1} \left|1+\frac{it}{N}\right|^{\alpha_1+\beta_1-\gamma_1-1} \exp\left(\frac{\pi}{2}\,|\alpha_2+\beta_2-\gamma_2|\right) \\
&\ll N^{\alpha_1+\beta_1-\gamma_1-1}\,\lambda_{\alpha_1+\beta_1-\gamma_1-1}(t),
\end{aligned}$$

where $\lambda_\nu(t)$ is defined by (8.144). Hence (8.150) yields

$$|J_N| \ll |z|^{N+\frac{1}{2}}\, N^{\alpha_1+\beta_1-\gamma_1-1} \int_{-\infty}^{+\infty} \lambda_{\alpha_1+\beta_1-\gamma_1-1}(t)\,\frac{e^{-\vartheta t}}{\cosh(\pi t)}\,dt. \tag{8.151}$$

Since $|\vartheta| < \pi$, the last integral converges. If $|z| < 1$, by (8.151) we get

$$\lim_{N\to+\infty} J_N = 0,$$

whence, making $N \to +\infty$ in (8.148), (8.149),

$$-\frac{1}{2\pi i}\int_{(k)} \frac{\Gamma(s+\alpha)\Gamma(s+\beta)}{\Gamma(s+\gamma)}\,\Gamma(-s)(-z)^s\,ds = -\frac{\Gamma(\alpha)\Gamma(\beta)}{\Gamma(\gamma)} \sum_{n=0}^{\infty} \frac{(\alpha)_n\,(\beta)_n}{(\gamma)_n}\,\frac{z^n}{n!}.$$

Thus we have proved (8.146) under the additional assumption $|z| < 1$. Since ${}_2F_1(\alpha, \beta; \gamma; z)$ is a regular function of z in $\mathbb{C}\backslash[1, +\infty)$ and the right-hand side of (8.146) is a regular function of z in $\mathbb{C}\backslash[0, +\infty)$, we conclude that (8.146) holds for all $z \in \mathbb{C}\backslash[0, +\infty)$ by analytic continuation. □

Bibliography

This short list is a selection of few among the numerous classical treatises dealing with the theory of functions of one complex variable and with special functions. These items have been selected with particular regard to the subjects treated in the present lecture notes.

G. E. Andrews, R. Askey, R. Roy. *Special Functions*, Encyclopedia of Mathematics and its Applications, vol. 71, Cambridge University Press, New York, 1999.

A. Erdélyi et al. *Higher Transcendental Functions*, 3 volumes, Bateman Manuscript Project, McGraw-Hill, New York, 1953.

H. Hochstadt. *The Functions of Mathematical Physics*, Dover Publications, New York, 1986.

E. L. Ince. *Ordinary Differential Equations*, Dover Publications, New York, 1956.

K. Knopp. *Theory of Functions*, parts I and II, translated by F. Bagemihl, Dover Publications, New York, 1996.

N. N. Lebedev. *Special Functions and their Applications*, translated by R. A. Silverman, Dover Publications, New York, 1972.

E. C. Titchmarsh. *The Theory of Functions*, 2nd edition, Oxford University Press, London, 1968.

E. T. Whittaker, G. N. Watson. *A Course of Modern Analysis*, 4th edition, Cambridge University Press, London, 1927.

C. Viola, *An Introduction to Special Functions*,
UNITEXT - La Matematica per il 3+2 102, DOI 10.1007/978-3-319-41345-7

Index

C. Viola, *An Introduction to Special Functions*,
UNITEXT - La Matematica per il 3+2 102, DOI 10.1007/978-3-319-41345-7

Printed in the United States
By Bookmasters